Primary SPACE Project Research Team

Research Co-ordinating Group

Professor Paul Black (Co-director)
Jonathan Osborne

Centre for Educational Studies
King's College London
University of London
Cornwall House Annexe
Waterloo Road
London SE1 8TZ

Tel: 071 872 3094

Dr Wynne Harlen (Co-director)
Terry Russell

Centre for Research in Primary Science
and Technology
Department of Education
University of Liverpool
126 Mount Pleasant
Liverpool L3 5SR

Tel: 051 794 3270

Project Researchers

Pamela Wadsworth (from 1989)

Derek Bell (from 1989)
Ken Longden (from 1989)
Adrian Hughes (1989)
Linda McGuigan (from 1989)
Dorothy Watt (1986-89)

Associated Researchers

John Meadows
(South Bank Polytechnic)

Bert Sorsby
John Entwistle
(Edge Hill College)

LEA Advisory Teachers

Maureen Smith (1986-89)
(ILEA)

Joan Boden
Karen Hartley
Kevin Cooney (1986-88)
(Knowsley)

Joyce Knaggs (1986-88)
Heather Scott (from 1989)
Ruth Morton (from 1989)
(Lancashire)

74

PRIMARY SPACE PROJECT
RESEARCH REPORT
January 1990

Evaporation and Condensation

by
TERRY RUSSELL and DOROTHY WATT

LIVERPOOL UNIVERSITY PRESS

First published 1990 by
Liverpool University Press
PO Box 147, Liverpool L69 3BX

Reprinted, with corrections, 1992

British Library Cataloguing in Publication Data
Data are available
ISBN 0 85323 446 9

Printed and bound by
Antony Rowe, Limited, Chippenham, England

CONTENTS

INTRODUCTION

This introduction is common to all SPACE topic reports and provides an overview of the Project and its programme of research.

The Primary SPACE Project is a classroom-based research project which aims to establish

- *the ideas which primary school children have in particular science concept areas*
- *the possibility of children modifying their ideas as the result of relevant experiences.*

The research is funded by the Nuffield Foundation and Unwin Hyman and is being conducted at two centres, the Centre for Research in Primary Science and Technology, Department of Education, University of Liverpool and the Centre for Educational Studies, King's College, London. The joint directors are Professor Wynne Harlen and Professor Paul Black. Three local education authorities are involved: Inner London Education Authority, Knowsley and Lancashire.

The Project is based on the view that children develop their ideas through the experiences they have. With this in mind, the Project has two main aims: firstly, to establish (through an elicitation phase) what specific ideas children have developed and what experiences might have led children to hold these views; and secondly, to see whether, within a normal classroom environment, it is possible to encourage a change in the ideas in a direction which will help children develop a more 'scientific' understanding of the topic (the intervention phase).

Eight concept areas have been studied so far:

Electricity
Evaporation and condensation
Everyday changes in non-living materials
Forces and their effect on movement
Growth
Light
Living things' sensitivity to their environment
Sound.

The Project has been run collaboratively between the University research teams, local education authorities and schools, with the participating teachers playing an active role in the development of the Project work.

Over the life-span of the Project a close relationship has been established between the University researchers and the teachers, resulting in the development of techniques which advance both classroom practice and research. These methods provide opportunities, within the classroom, for children to express their ideas and to develop their thinking with the guidance of the teacher, and also help researchers towards a better understanding of children's thinking.

The involvement of teachers

Schools and teachers were not selected for the Project on the basis of a particular background or expertise in primary science. In the majority of cases, two teachers per school were involved, which was advantageous in providing mutual support. Where possible, the Authority provided supply cover for the teachers so that they could attend Project sessions for preparation, training and discussion during the school day. Sessions were also held in the teachers' own time, after school.

The Project team aimed to have as much contact as possible with the teachers throughout the work to facilitate the provision of both training and support. The diversity of experience and differences in teaching style which the teachers brought with them to the Project meant that achieving a uniform style of presentation in all classrooms would not have been possible, or even desirable. Teachers were encouraged to incorporate the Project work into their existing classroom organisation so that both they and the children were as much at ease with the work as with any other classroom experience.

The involvement of children

The Project involved a cross-section of classes of children throughout the primary age range. A large component of the Project work was classroom-based, and all of the children in the participating classes were involved as far as possible. Small groups of children and individuals were selected for additional activities or interviews to facilitate more detailed discussion of their thinking.

The structure of the Project

For each of the concept areas studied, a list of concepts was compiled to be used by researchers as the basis for the development of work in that area. These lists were drawn up from the standpoint of accepted scientific understanding and contained concepts which were considered to be a necessary part of a scientific understanding of each topic. The lists were not necessarily considered to be statements of the understanding which would be desirable in a child at age eleven, at the end of the Primary phase of schooling. The concept lists defined and outlined the area of interest for each of the studies; what ideas children were able to develop was a matter for empirical investigation.

Most of the Project research work can be regarded as being organised into four phases, preceded by an extensive pilot phase. These phases are described in the following paragraphs and are as follows:

Pilot work
Phase 1: Exploration
Phase 2: Pre-Intervention Elicitation
Phase 3: Intervention
Phase 4: Post-Intervention Elicitation

The phases of the research

Each phase, particularly the Pilot work, was regarded as developmental; techniques and procedures were modified in the light of experience. The modifications involved a refinement of both the exposure materials and the techniques used to elicit ideas. This flexibility allowed the Project team to respond to unexpected situations and to incorporate useful developments into the programme.

There were three main aims of the Pilot phase. Firstly, to trial the techniques used to establish children's ideas; secondly, to establish the range of ideas held by primary school children; and thirdly, to familiarise the teachers with the classroom techniques being employed by the Project. This third aim was very important since teachers were being asked to operate in a manner which, to many of them, was very different from their usual style. By allowing teachers a 'practice run', their initial apprehensions were reduced, and the Project rationale became more familiar. In other words, teachers were being given the opportunity to incorporate Project techniques into their teaching, rather than having them imposed upon them.

In the Exploration phase children engaged with activities set up in the classroom for them to use, without any direct teaching. The activities were designed to ensure that a range of fairly common experiences (with which children might well be familiar from their everyday lives) was uniformly accessible to all children to provide a focus for their thoughts. In this way, the classroom activities were to help children articulate existing ideas rather than to provide them with novel experiences which would need to be interpreted.

Each of the topics studied raised some unique issues of technique and these distinctions led to the Exploration phase receiving differential emphasis. Topics in which the central concepts involved long-term, gradual changes, e.g. 'Growth', necessitated the incorporation of a lengthy exposure period in the study. A much shorter period of exposure, directly prior to elicitation was used with 'Light' and 'Electricity', two topics involving 'instant' changes.

During the Exploration, teachers were encouraged to collect their children's ideas using informal classroom techniques. These techniques were:

i. *Using log-books (free writing/drawing)*

Where the concept area involved long-term changes, it was suggested that children should make regular observations of the materials, with the frequency of these depending on the rate of change. The log-books could be pictorial or written, depending on the age of the children involved, and any entries could be supplemented by teacher comment if the children's thoughts needed explaining more fully. The main purposes of these log-books were to focus attention on the activities and to provide an informal record of the children's observations and ideas.

ii. *Structured writing/drawing*

Writing or drawings produced in response to a particular question were extremely informative. This was particularly so when the teacher asked children to clarify their diagrams and themselves added explanatory notes and comments where necessary, after seeking clarification from children.

Teachers were encouraged to note down any comments which emerged during dialogue, rather than ask children to write them down themselves. It was felt that this technique would remove a pressure from children which might otherwise have inhibited the expression of their thoughts.

iii. *Completing a picture*

Children were asked to add the relevant points to a picture. This technique ensured that children answered the question posed by the Project team and reduced the possible effects of competence in drawing skills on ease of expression of ideas. The structured drawing provided valuable opportunities for teachers to talk to individual children and to build up a picture of each child's understanding.

iv. *Individual discussion*

It was suggested that teachers use an open-ended questioning style with their children. The value of listening to what children said, and of respecting their responses, was emphasised as was the importance of clarifying the meaning of words children used. This style of questioning caused some teachers to be concerned that, by accepting any response whether right or wrong, they might implicitly be reinforcing incorrect ideas. The notion of ideas being acceptable and yet provisional until tested was at the heart of the Project. Where this philosophy was a novelty, some conflict was understandable.

In the Elicitation phase, the Project team collected structured data through individual interviews and work with small groups. The individual interviews were held with a random, stratified sample of children to establish the frequencies of ideas held. The same sample of children was interviewed pre- and post-Intervention so that any shifts in ideas could be identified.

The Elicitation phase produced a wealth of different ideas from children, and led to some tentative insights into experiences which could have led to the genesis of some of these ideas. During the Intervention teachers used this information as a starting point for classroom activities, or interventions, which were intended to lead to children extending their ideas. In schools where a significant level of teacher involvement was possible, teachers were provided with a general framework to guide their structuring of classroom activities appropriate to their class. Where opportunities for exposing teachers to Project techniques were more limited, teachers were given a package of activities which had been developed by the Project team.

Both the framework and the Intervention activities were developed as a result of preliminary analysis of the Pre-Intervention Elicitation data. The Intervention strategies were:

(a) Encouraging children to test their ideas

It was felt that, if pupils were provided with the opportunity to test their ideas in a scientific way, they might find some of their ideas to be unsatisfying. This might encourage the children to develop their thinking in a way compatible with greater scientific competence.

(b) Encouraging children to develop more specific definitions for particular key words

Teachers asked children to make collections of objects which exemplified particular words, thus enabling children to define words in a relevant context, through using them.

(c) Encouraging children to generalise from one specific context to others through discussion.

Many ideas which children held appeared to be context-specific. Teachers provided children with opportunities to share ideas and experiences so that they might be enabled to broaden the range of contexts in which their ideas applied.

(d) Finding ways to make imperceptible changes perceptible

Long-term, gradual changes in objects which could not readily be perceived were problematic for many children. Teachers endeavoured to find appropriate ways of making these changes perceptible. For example, the fact that a liquid could 'disappear' visually and yet still be sensed by the sense of smell - as in the case of perfume - might make the concept of evaporation more accessible to children.

(e) Testing the 'right' idea alongside the children's own ideas

Children were given activities which involved solving a problem. To complete the activity, a scientific idea had to be applied correctly, thus challenging the child's notion. This confrontation might help children to develop a more scientific idea.

In the Post-Intervention Elicitation phase the Project team collected a complementary set of data to that from the Pre-Intervention Elicitation by re-interviewing the same sample of children. The data were analysed to identify changes in ideas across the sample as a whole and also in individual children.

These four phases of Project work form a coherent package which provides opportunities for children to explore and develop their scientific understanding as a part of classroom activity, and enables researchers to come nearer to establishing what conceptual development it is possible to encourage within the classroom and the most effective strategies for its encouragement.

The implications of the research

The SPACE Project has developed a programme which has raised many issues in addition to those of identifying and changing children's ideas in a classroom context. The question of teacher and pupil involvement in such work has become an important part of the Project, and the acknowledgement of the complex interactions inherent in the classroom has led to findings which report changes in teacher and pupil attitudes as well as in ideas. Consequently, the central core of activity, with its pre- and post-test design, should be viewed as just one of the several kinds of change upon which the efficacy of the Project must be judged.

The following pages provide a detailed account of the development of the Evaporation and Condensation topic, the Project findings and the implications which they raise for science education.

The research reported in this and the companion research reports, as well as being of intrinsic interest, informed the writing and development with teachers of the Primary SPACE Project curriculum materials, published by Unwin Hyman.

1. METHODOLOGY

1.1 Sample

a. Schools

Six schools (three primary, two junior and one infant) within the Knowsley LEA were involved in this topic. Between the six schools, every age-group from five to eleven years was represented. Twelve teachers participated in the Project work, two each from three of the schools, four from one school and one each from the remaining two schools.

A number of staffing changes occurred during the course of the Project in the six schools involved, resulting in the loss of three teachers in July 1987, all of whom were replaced. In one case two other teachers from the same school joined in. In the other case a school withdrew from the Project and a second teacher at an existing SPACE school became involved.

Names of the participating teachers, their schools and head teachers are in Appendix 1.

b. Teachers

Teachers were selected to participate in the Project by their LEA, and this selection was not on the basis of any particular background or expertise in Science. In the latter part of the Project the LEA provided supply cover so that teachers could attend meetings during the school day to prepare them for Project work. Prior to the supply cover being available teachers met for regular after-school meetings at the research team's offices.

c. Children

All children in the classes of the participating teachers were involved in the Project work to some extent. Additionally, a stratified random sample of children was interviewed from each of the twelve participating classes. This sample was balanced for age, sex and achievement. This third measure was the class teacher's subjective rating about whether a child was high, middle or low on overall school achievement.

d. Liaison

The University researchers were supported by the team of three Knowsley ESG Primary Science teachers who visited teachers in their schools, providing support and guidance, and also carrying out some individual interviews. These three Group Co-ordinators provided a valuable link between the teachers and the University.

1.2 The Research Programme

Classroom work concerning 'Evaporation and Condensation' took place at three different points during the school year. Each of these periods of classroom activity was followed by a phase of one-to-one interviewing by members of the extended research team. The phases were as follows:

Pilot (March 1987)

Interviews (April 1987)

Exploration (October 1987)

Interviews (October 1987)

Intervention (November 1987)

Interviews (January 1988)

Defining 'Evaporation/Condensation'

A list of concepts concerning the phenomena of evaporation and condensation was compiled to provide a framework for work in these areas:

1. Water can exist in the form of water vapour, an invisible, odourless gas.

2. Air contains this invisible vapour.

3. The amount of water vapour present in the air can vary.

4. Evaporation and condensation separate water from solids which are dissolved in it.

5. Other liquids also evaporate and condense.

6. Evaporation leads to cooling. Cooling leads to condensation.

2. PRE-INTERVENTION ELICITATION WORK

2.1 Pilot Phase (March 1987)

Teachers were given a pack containing information about the Project and outlines of techniques and activities to enable them to cary out Pilot work with their class. This pack may be found as Appendix II.

Activities

The activities which were put into the classrooms for the five-week Pilot period were as follows:

Evaporation: drying out of sliced bread
monitoring the water level in a large container of water
monitoring the water level of solutions
clothes drying
monitoring the level of water in puddles
drying-up of water-based paints

Condensation: breath on windows
breath in cold air

The following paragraphs detail what was involved in each activity, and give an indication of the reactions of pupils and teachers to them:

a. Drying-out of bread

A fresh slice of bread was drawn around before being left out for a day. It was then drawn around again, and fresh and dried-up slices compared. The changes in size and texture of the bread slice as the water evaporated from it were very noticeable and provoked plenty of interest. However, there was very little mention of water loss as an explanation for the changes, the bread being described as having 'gone stale', 'dried up' or 'been left out'. It was felt firstly that many children may be unaware that bread contains moisture and secondly that the familiar nature of the change could be accounted for by children in everyday descriptive terms.

b. Monitoring the water level in a large container of water

A water tank, bucket or other transparent container which enabled a large surface area of water to be exposed to the air was filled with water and the level of the water was marked. Over five weeks, the water level was monitored periodically, the interval between observations being dependent upon the rate of evaporation. This very simple activity was extremely effective in eliciting ideas from children about where they thought the water had gone. While this activity might be anticipated to be of little intrinsic interest, it forced attention to be focused upon the phenomenon of evaporation and this happening proved puzzling and thought-provoking for the children.

c. Monitoring the water level of solutions

Small amounts of sugar, salt and coffee were each dissolved in a saucer full of water and the level of the solution marked. These containers were observed at intervals until all the water had evaporated. These activities also provoked thought and interest in many children though the additional phenomenon of a 'solution' caused some confusion. It appeared that some children associated 'coffee' with the solution which is drunk rather than the granules or powder from which the solution is made. This is another instance of everyday parlance leading to ambiguity in the interpretation of children's responses.

d. Clothes drying

A small number of clothes, or pieces of material, was thoroughly wetted in water and hung on a line to dry. This line was either inside or outside the classroom depending on the weather and on the location of the particular classrooms within the schools. This activity, very similar to what children were likely to have experienced at home, was interesting for the children since many of them were actively involved in setting it up. It fitted well into a 'science' lesson and so was easily managed by the teachers.

e. Monitoring the level of water in puddles

Artificial puddles were made on the school playground using tap water and polythene sheeting and watched during the day. This activity fitted into a school day and was organisationally unproblematic for the teachers. The children enjoyed this activity because they were able to predict outcomes from their previous experience with rainwater puddles.

f. Drying-up of paints

A paint palette had some water-based paint left in it to dry out. The children were asked whether they could make the paint useable again, and the overwhelming response was, "Yes; add water". However, there seemed to be little understanding of the process of paint drying, and water was not often mentioned in connection with the drying-up of the paints.

g. Breath on windows

Children were asked to observe what happened when they breathed out onto a window until the 'breath' was no longer visible. The condensation produced in this way was always easily noticeable and was commented on by children though with little reference to water.

h. Breath in cold air

This activity needed suitable weather conditions, but this did not seem to be a problem because of the time of year at which it was carried out. The condensation was easily noticeable though water was referred to infrequently.

Despite the relatively large number of activities there did not appear to be too much of a problem fitting them into the classroom schedules. Those activities in which changes were gradual and long-term, eg. the large container of water, the solutions and the paints drying, were quickly set up and could be left unobtrusively while still provoking thought and interest in the children. The other, shorter activities could be timetabled as science lessons and were easily managed in that way. An additional advantage of all these activities was that the necessary equipment, eg. saucers, paint, bowls and water was readily available in the schools.

Elicitation techniques

During this Pilot phase each child was given an individual 'diary' by their teacher. These notebooks were not used for daily recording of observations but tended to be used in conjunction with the setting up of particular activities. Since many of the activities lasted for a day or less, teachers often asked their classes to write about what they had just done and seen and what they thought had happened, during convenient lulls in the activity, (e.g. when the clothes were drying), and at its close. For recording ideas concerned with the longer term activities, teachers found that a suitable time interval between observations was about a week. The children then made a weekly entry in their books to correspond with each activity. These 'diaries' were found to be an effective way of finding out children's ideas, particularly if the writing was accompanied by a diagram or picture.

Some teachers of younger children found that an attempt to produce diaries was unproductive since children's recording skills were not sufficiently developed to allow them to put their thoughts onto paper. These teachers relied heavily on tape-recording group and individual discussion as a means of logging children's ideas. This method of recording led to the production of verbatim transcripts of children's ideas which, in some cases, were more illuminating than the diaries produced by older children whose abilities to record thoughts clearly were not fully matured.

After three weeks of Pilot exposure activities, a sample of children from each participating class was selected to be interviewed individually by a member of the Project team. The selection on this occasion was on the basis of the ideas the child had expressed during the classroom activities so that a wide range of ideas could be probed fully in one-to-one discussion.

Children were interviewed about evaporation and condensation using phenomena which were slightly different from those experienced in their classrooms. These examples were selected from:

evaporation:
- handprints made on a paper towel and shaken dry
- hands wetted and shaken dry
- a newly painted picture drying

condensation:
- the outside of a tin containing iced water
- breath on a mirror

The interviewer had a series of questions relating to each activity around which to base the interview, though the list was not considered to be exclusive if further questions were necessary to clarify answers. The interview questions are presented as Appendix III.

Reactions of Teachers to Project Work

The Pilot phase was very successful within the classroom in terms of providing activities which supported the effective elicitation of children's ideas about evaporation. The activities were felt to be manageable within the classroom though the same ideas could have been elicited by using far fewer activities since there was some degree of redundancy. In this respect, it was interesting to note that many teachers remarked upon their children's surprise at being asked similar questions about each activity when many of the class did not see the activities as inter-related. Additionally, several activities seemed to be so 'everyday' that many children appeared not to think about explaining them beyond describing what happened, e.g. 'It dried'. This non-explanation might imply an acceptance that certain events just happen and, while this in itself is of interest, the activities such as paint and bread drying out provided little insight into children's understanding. These activities were not taken beyond the Pilot phase. The Pilot phase also enabled teachers to become more familiar with the technique of open questioning and non-directive elicitation within their classrooms, though several had expressed anxieties about not providing the children with answers to questions. The increasing realisation that the children themselves had ideas to share enabled the teachers to respond constructively to this phase of Project work, and to make suggestions concerning the way this work could effectively be handled in the classroom.

2.2 Exploration (October 1987)

The overriding aim of the Exploration phase proper was to provide a setting for the elicitation of children's ideas within the classroom. The issues raised by teachers as a result of the Pilot work, and also the relative merits of particular activities in leading to the successful elicitation of ideas were therefore given careful consideration in the development of the forthcoming Exploration.

The structure of the Exploration period was revised considerably to take into account:

a. refinement of activities for use within a shorter time span
b. developments in thinking about classroom elicitation techniques
c. prerequisites of the Intervention phase, which would follow the elicitation.

a. A shorter time span

The five-week Exploration period during the Pilot work had the advantages of allowing a considerable degree of change to occur in the activities which involved only a slow evaporation of water. However, while these activities provoked thought it was felt that a shorter period would allow this to be done just as effectively since the evaporation was not dramatic enough to sustain the interest of children for a considerable length of time. The Exploration was shortened to two weeks prior to interviewing, a period which was long enough for noticeable changes to occur in the activities and for children to reflect on the phenomenon. The number of activities was also reduced to remove redundancies and to allow classroom sessions to be spaced out sufficiently within the two-week period.

b. Classroom elicitation techniques

The manner in which the diaries had been used during the Pilot phase provided an insight into possible developments of the technique which might improve the quality of responses the children gave, taking them beyond description. The individual diaries, which had proved very helpful in the Pilot work, had not actually been used as diaries except in connection with one long-term change, that of evaporation from the large container of water; all other entries had been one-off descriptive accounts of the shorter activities. It was felt that, by posing specific questions which could be responded to diagrammatically or in writing, children might be encouraged to express their ideas succinctly without giving a full written account of their activity or simply recording observations.

It was suggested that the individual diaries should be replaced by a class log-book situated next to the large container of water so that children could record their thoughts and observations as and when they wanted to. It was also felt that the group or class log-book would help to keep the activities incidental within the classroom, with the attendant advantage that children would make contributions when they felt it was relevant to do so.

c. Forthcoming Intervention

The rationale behind the Project required that children should test out their own ideas to encourage them to develop their thinking. It was therefore important that the activities used to elicit the ideas had scope for extension so they could be modified as part of an investigation of the phenomenon of evaporation, e.g. if the heat was suggested to make the water in the tank go then it would be possible for the tank to be placed somewhere away from heat if that was the test the child suggested.

It was hoped that posing questions to which children responded with diagrams would allow teachers to have access, in an easily assessable format, to a number of their children's ideas. An early awareness of prevailing ideas meant that teachers would be prepared for the Intervention phase when the children would be working with their ideas. A checklist was also developed from the emergent ideas in the Pilot phase to aid teachers in building up a picture of the frequency with which various ideas were held by children in their classes.

Activities

As in the Pilot phase, teachers were requested to proceed with a light touch and to keep the elicitation work as incidental as possible. The importance of open questioning and of maintaining a non-didactic stance was emphasised. It was suggested that the meanings of certain key words, e.g. 'dry' could be probed. Teachers were again provided with a detailed description of each activity, and also with a checklist to assist their recording of children's ideas. This checklist was derived from the categorisation of children's ideas from the Pilot phase. It was emphasised that it should not be used as a question and answer sheet, but simply as an aid to teacher assessment of the children's ideas. These teacher guidelines are in Appendix IV.

The activities set up in the classroom were:

evaporation:
- . monitoring the water level in a large container of water
- . monitoring the water level of solutions
- . clothes drying

condensation:
- . breath in cold air

These activities have already been described as part of the Pilot phase, and the materials used were very similar this second time around, though with important differences in the manner in which children's attention was focused.

Elicitation Techniques

a. Labelled diagrams

Three tasks were set, which teachers were asked to introduce in as informal a way as possible so that they did not appear to be tests. It was suggested that each task be done when the related activity was at a certain stage so that the children had had the opportunity to experience the relevant phenomenon. The tasks were as follows:

1. Evaporation of 'Pure' Water

   Draw a picture of/write about the water tank (or similar container) and show

   i) where you think the water has gone
   ii) how you think the water got there.

   Do you think you could make the water in the tank last longer/go faster?

   (drawn after the tank had been set up for two weeks).

2. Clothes Drying

   Draw a picture showing/write about where you think would be the best place to put the clothes/materials so that they will dry as quickly as possible. Also show what you think will happen to the water.

   (drawn after the clothes/material had been washed but before they dried).

3. Condensation of Breath in Air

   Describe (in words or pictures) what you think is happening. Where do you think it comes from? Do you think you can make it go away?

   (done directly following breathing in cold air).

This technique proved to be a very valuable method of establishing a child's ideas. Its value was enhanced by teacher annotation which was possible since children drawing make relatively low demands on direct teacher supervision. Drawings produced by infant children needed more extensive clarification since their recording skills were more limited but the results were still valuable, though less predictable.

b. Class log-book

The log-book was not used with the same confidence by children as the individual diaries had been. Several teachers found that children were reluctant to make any entries, even observational, so they gave children individual sheets of paper to write on and then stuck them into the log-book. This approach worked successfully though it gave the activity more prominence and placed more onus on the child to respond.

c. Individual discussion

Teachers appeared to find individual discussion increasingly useful, possibly as their ease with open questioning methods increased. The diagram tasks provided teachers with a framework within which to question children and this focus was helpful. However, more extended questioning was very time-consuming and not often possible within the classroom set-up.

Organisation

The Project activities were interesting for the children and easily handled within a classroom situation. Teachers expressed the desire to encourage all children to put forward their views and felt that the diagram tasks had been more successful in this respect than the class log-book. The main concern of some of the teachers was the amount of time they wanted to spend with individuals and small groups of children within the school day. There were classrooms in which there was little precedent for small group discussion and where attempts to initiate it led teachers into organisational difficulties.

2.3 Elicitation - Individual Interviews

A random, stratified sample of children was selected from each class to be interviewed individually by a member of the Project team. The sample was balanced by gender and achievement band. The purpose of these individual interviews was quite different from the Pilot phase. In the Pilot phase, the interviews served to elaborate and clarify the qualitative range of ideas which had been established by teachers during classroom activities. The purpose of moving towards a random sample was to enable the Project to comment on the relative frequency of the main ideas which children were producing. Any shifts in children's thinking after the Intervention period might be described, for groups and for individuals.

The number of activities in the Exploration phase was small enough to allow children to be interviewed about each one. Two activities from the Pilot phase interviews (handprints on paper towels and condensation on a tin of iced water) were also used. These two interview activities were included to give children experiences of the phenomena actually during the interview so that children were not being asked to respond solely from memory.

The interviewer had a set of questions around which to base the interview. These questions were similar to those used during the Pilot phase, and may be found in Appendix V. The length of the interview varied greatly from child to child but was on average 30-45 minutes.

Reactions of Teachers to Project Work

Largely due to the experience gained during the Pilot phase, the Exploration period proved very successful within the classrooms, allowing the children to express their ideas in a structured way. As a result of this phase of work teachers were able to summarise very succinctly the ideas emergent in their classes, and to discover similarities between their own and other teacher's children when experiences were exchanged at group meetings.

3. CHILDREN'S IDEAS

An informal look at children's ideas

The children from each class produced a large amount of work in connection with the Pilot phase and Exploration period. This work was requested partly to ensure that the 'thinking activities' did not become ignored and recede into the background in the classroom, and partly to help children to articulate their thoughts. The classwork was also important in providing teachers with access to their class's ideas so they could begin to consider how to approach the Intervention work. The classroom work, in the form of writing, drawings and teacher notes from discussions, has been summarised below as fully as its varying methods of presentation would allow. It is necessary to confront the possibility that these expressions of ideas on paper may represent only the ideas that a child is able to communicate, rather than the comprehensive, relatively unambiguous account of the sort that is possible during one-to-one interviewing. Teachers have made efforts to clarify ambiguities with annotation, though this was not possible in every case.

This section contains a descriptive account of children's responses given in answer to questions about three activities: monitoring the water level of a large container of water (hereafter referred to as the 'water tank'); wetting and drying some material; condensation of breath in cold air, or against a window. Where percentage figures are given these were obtained by categorising all of the annotated diagrams and writing relating to the three activities. All of the children in the classes participating in Project work were included in the sample.

The children's work was categorised according to responses to three questions:

. where has the water gone?	(final location)
. what has made the water go?	(agent and conditions)
. can the water be made to go faster/slower?	(accelerating and retarding evaporation)

The categories were generated by the children's responses and frequency counts were then made of particular ideas.

Where has the water gone?

The range of children's responses to this question was wide. However, considering all three activities, it is possible to describe the responses with reference to a common set of categories. While certain categories appear to be more relevant to particular activities, most are applicable to all three.

The condensation activity, while not being recognised as involving water by many children, involved evaporation when the condensate evaporated and so is included here. Questions asked of children in this latter context avoided mentioning water by asking 'Where has it gone?'.

a. The water tank

A very common response from young children was that the water 'dried up' or 'went down' (45% of infants) with no further indication of where the water had gone. This suggests that children were aware that the amount of water had diminished in the tank, but that the whereabouts of the missing water was unknown to them. In fact, some children seemed not to need to explain the missing water - it had gone and therefore did not exist. This concentration on the remaining water in the water tank became less prevalent as children got older (21% of lower juniors, 5% of upper juniors). This age-related shift in response might be due in part to the children being able to conserve volume.

Fig. 3.1

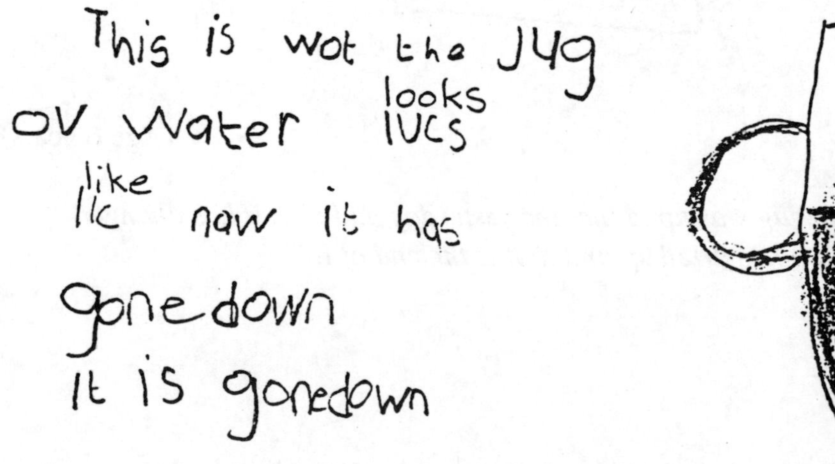

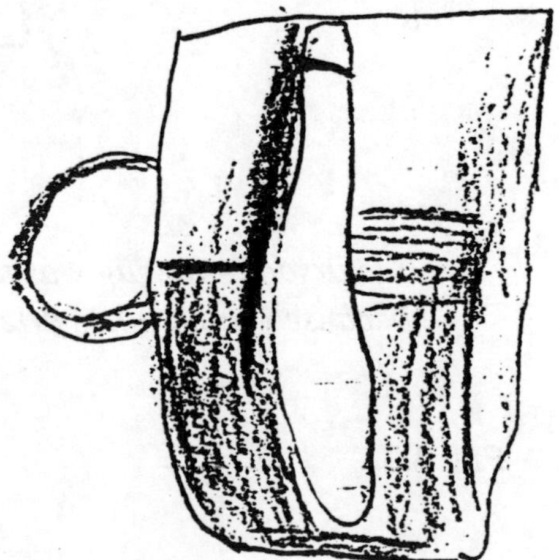

(Age 6 years)

"This is what the jug of water looks like now it has gone down. It is gone down."

Fig 3.2

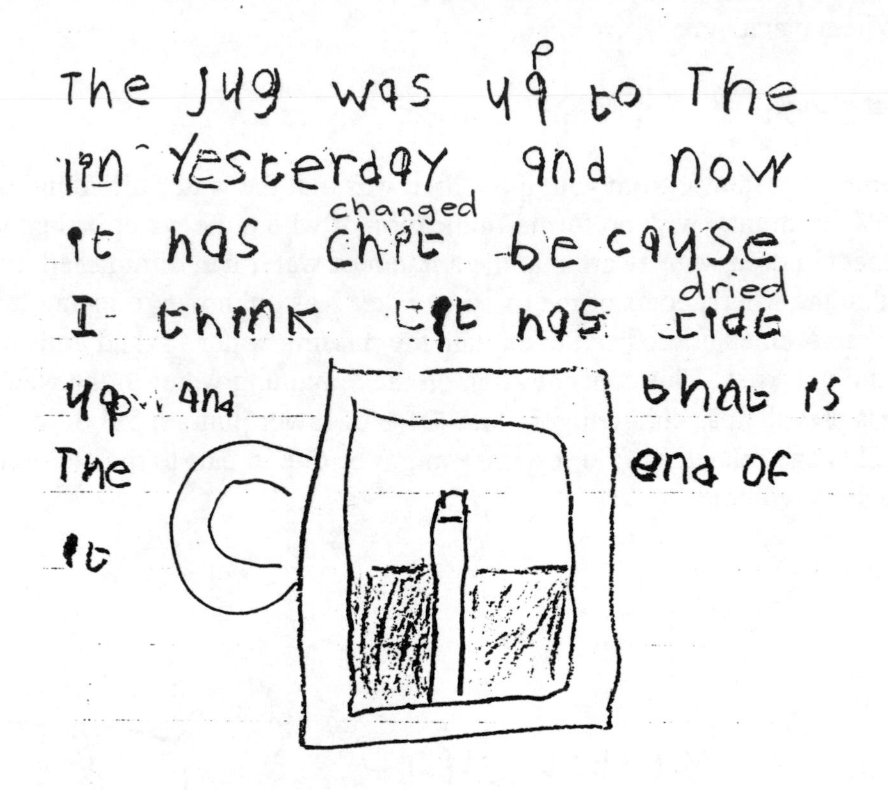

(Age 6 years)

"Thursday - the jug was up to the line yesterday and now it has changed because I think it has dried up and that is the end of it."

Fig. 3.3

the water has evaprated and in the week end
it went down.

(Age 8 years)

"the water has evaporated and in the weekend it went down."

Fig. 3.4

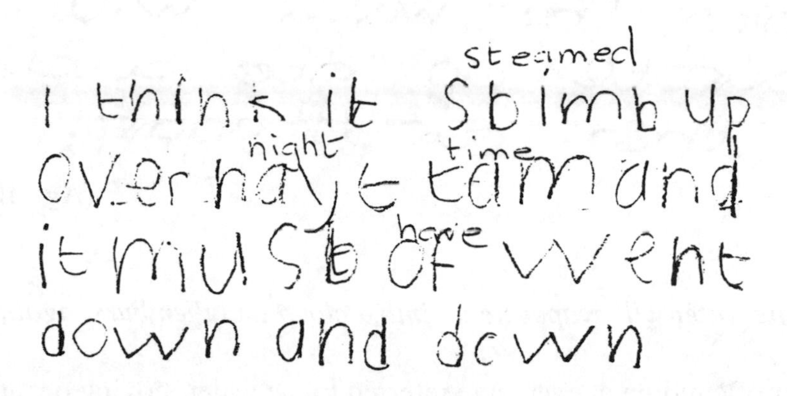

(Age 6 years)

"I think it steamed up over night time and it must have went down and down."

Words such as 'disappeared' and 'evaporated' are used with reference to the water with increasing frequency as the pupils get older. While increasing use of 'scientific' terms connected with evaporation might be expected the use of words like 'disappeared' is surprising. It is possible to speculate that these words, when accompanied by no further explanation, do not simply reflect a greater facility with words on the part of the older children but perhaps indicate initial attempts to express a change in state of the observable water. This suggestion is supported by the association of such terms with mention of the air, clouds or sun as either an agent causing evaporation or the new location for the evaporated water.

Fig. 3.5

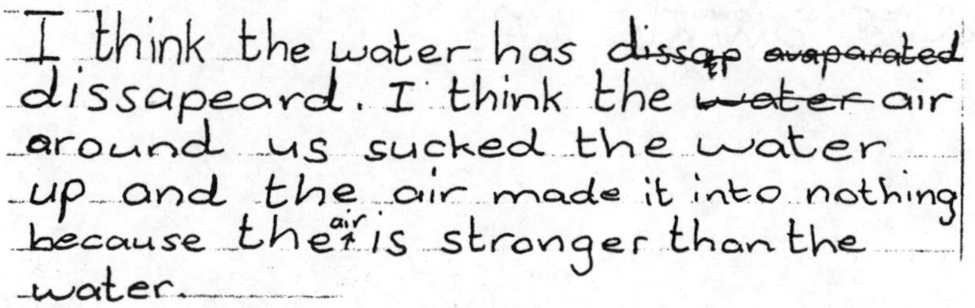

I think the water has ~~dissap~~ ~~evaporated~~
dissapeard. I think the ~~water~~ air
around us sucked the water
up and the air made it into nothing
because the air is stronger than the
water.

(Age 9 years)

"I think the water has disappeared. I think the air around us sucked the water up and the air made it into nothing because the air is stronger than the water."

Fig. 3.6

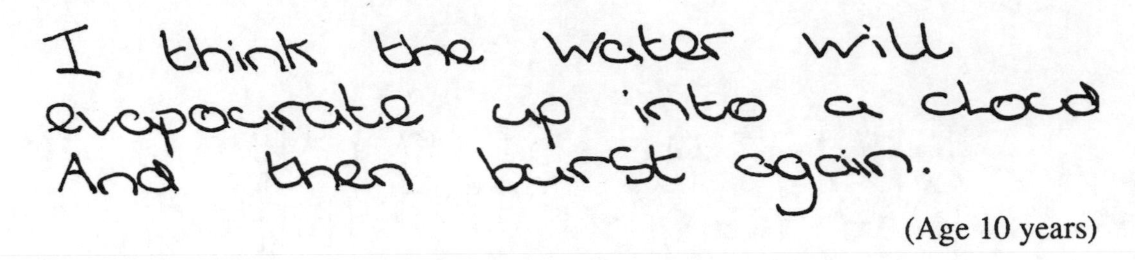

(Age 10 years)

"I think the water will evaporate up into a cloud and then burst again."

A small number of children at every age referred to the water moving or running to some nearby location, either into or under the tank the water was in or through holes in the water tank onto the table or floor, or to the drains or other part of the water supply system. In order to give this response, children had to be aware that volume was conserved and so they postulated something happening which would remove the water from sight and somehow move it to its new location. It was, however, water in its liquid state that was thought to be moved.

Fig. 3.7

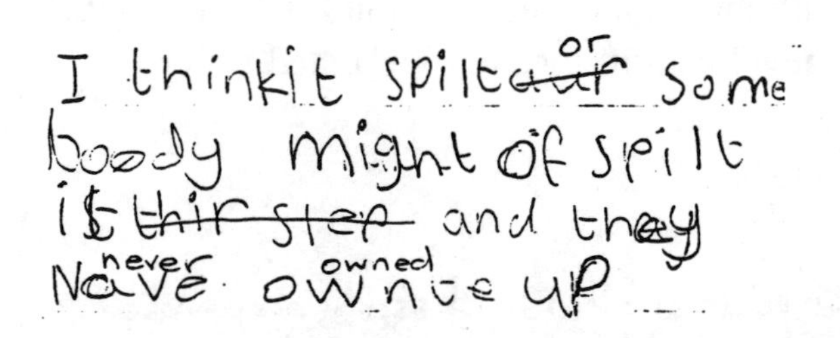

(Age 6 years)

"I think it spilt or somebody might have spilt it and they never owned up."

Fig. 3.8

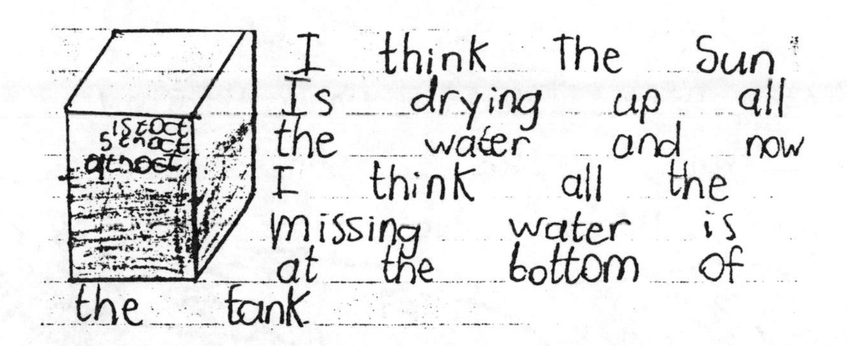

(Age 7 years)

"I think the sun is drying up all the water and now I think all the missing water is at the bottom of the tank."

Some children, sure that the water must have moved, went as far as searching for it to confirm their idea.

Fig. 3.9

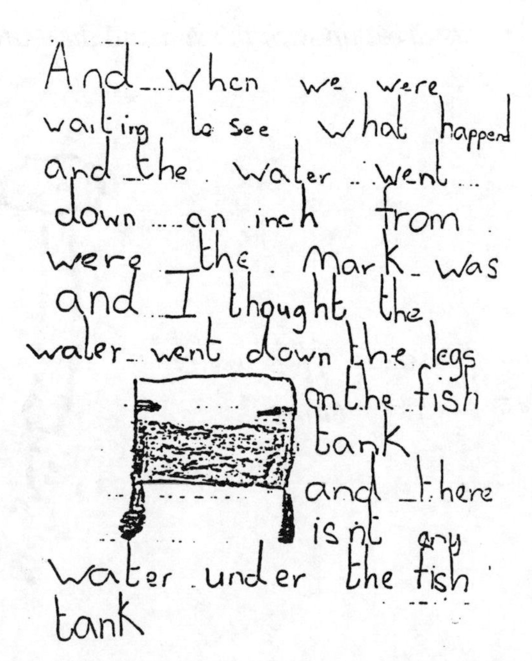

(Age 6 years)

"And when we were waiting to see what happened and water went down an inch from where the mark was and I thought the water went down the legs on the fish tank and there isn't any water under the fish tank."

Another group of children thought the water had gone to a more remote location, for example the sky, clouds or sun; this usually involved some causal reasoning.

Fig. 3.10

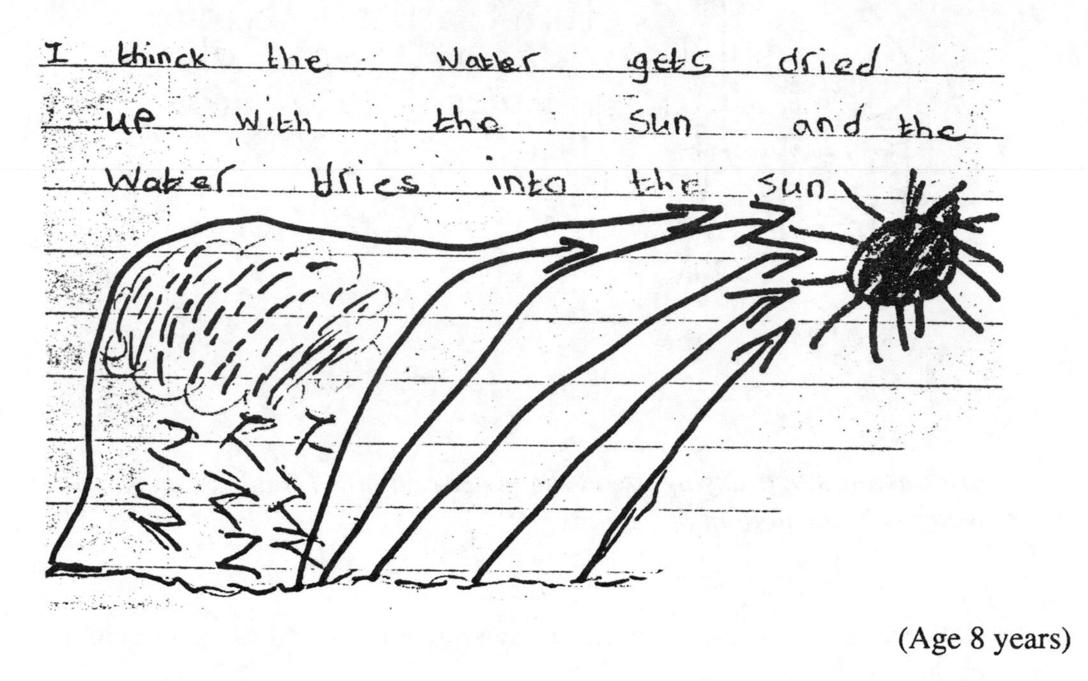

(Age 8 years)

"I think the water gets dried up with the sun and the water dries into the sun."

Fig. 3.11

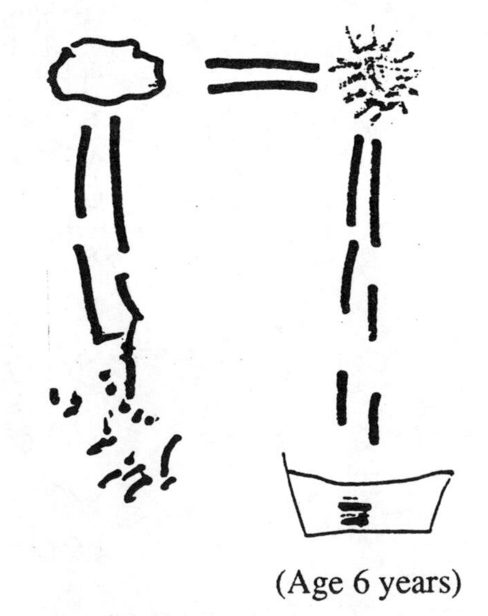

(Age 6 years)

"The sun passes the water to the clouds and it rains."

Fig. 3.12

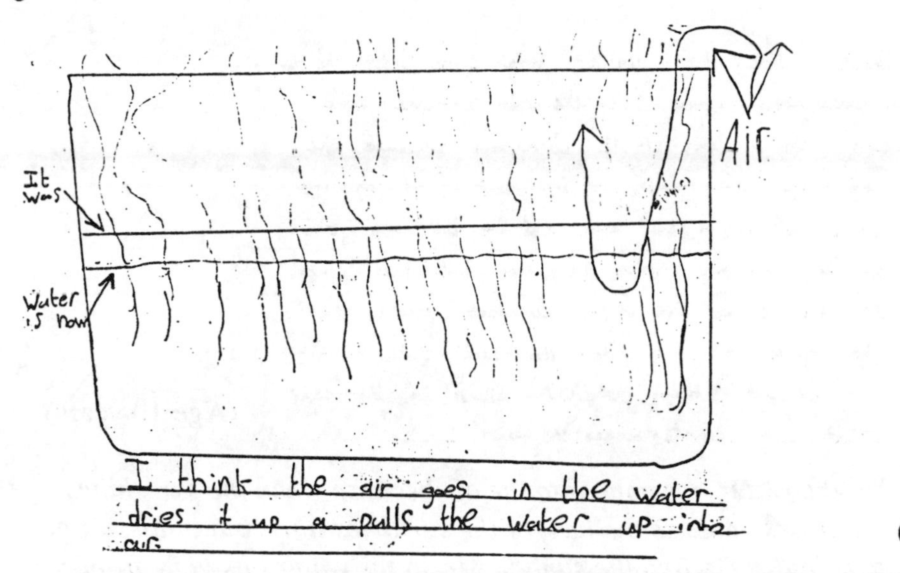

(Age 8 years)

"I think the air goes in the water dries it up and pulls the water up into air."

Taking all three distant locations together, there is a clear trend towards the older junior children being more likely to have suggested one of these locations (18% of infants, 38% of lower juniors, 65% of upper juniors). However, looking at each of the three separately some interesting patterns emerge. Clouds were mentioned by all ages with a doubling in frequency in the upper juniors. The air/sky, on the other hand was not mentioned at all by infant children, by 22% of lower juniors and then by 42% of upper juniors. This dramatic increase could possibly be explained in terms of the air itself being a phenomenon which was difficult for young children to understand. The clouds, on the other hand, are plainly visible and may easily be associated with rain. A different trend involves mention of the sun as the receptor of the water. The sun, like the clouds, was mentioned by children of all ages but the most frequent mention was in the lower juniors (19%) with a rapid tail-off in the upper juniors (2%). While it is scientifically accurate to suggest either the air or clouds as water-carriers the same is not true for the sun. A reason for this response might be that the sun had, through children's experiences of evaporation, been found to be a more obvious causal agent in the process of evaporation. The sun might therefore be considered to be an 'intermediate receptacle' for the water before children's understanding of the process of evaporation had developed sufficiently to include notions of water transformation which would lead more naturally to the air and clouds holding the water. This intermediate idea could provide a coherence between agency and location which would be helpful to children, and the notion appears to have been elaborated by some children to a high degree of complexity.

Fig. 3.13

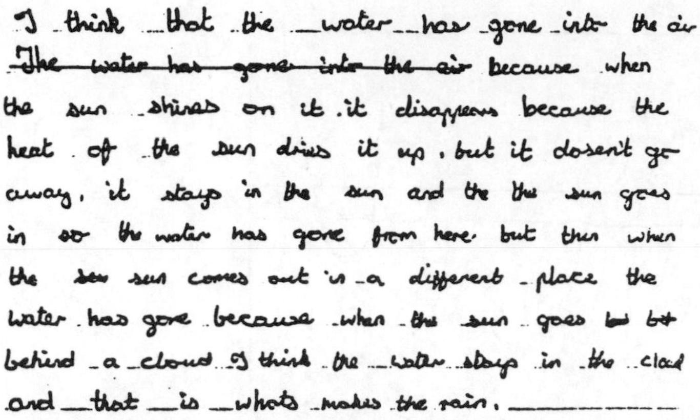

I think that the water has gone into the air
~~The water has gone into the air~~ because when
the sun shines on it. it disappears because the
heat of the sun dries it up. but it dosen't go
away. it stays in the sun and the the sun goes
in so the water has gone from here. but then when
the sew sun comes out is a different place the
water has gone because when the sun goes bt bt
behind a cloud I think the water stays in the clad
and that is whats makes the rain.

(Age 10 years)

"I think that the water has gone into the air because when the sun shines on it, it disappears because the heat of the sun dries it up, but it doesn't go away, it stays in the sun and the sun goes in so the water has gone from here but this when the sun comes out is a different place. The water has gone because when the sun goes behind a cloud I think the water stays in the cloud and that is what makes the rain."

The peak in 'sun' responses with the lower juniors coincides with a slight trough in 'cloud' responses. This might again be due to children finding the coherence of the sun aiding evaporation and storing the water more helpful and easier to envisage than the sun enabling water to reach the clouds. Some of the vocabulary that children used to describe the mechanism of evaporation suggests that this might be the case.

Fig. 3.14

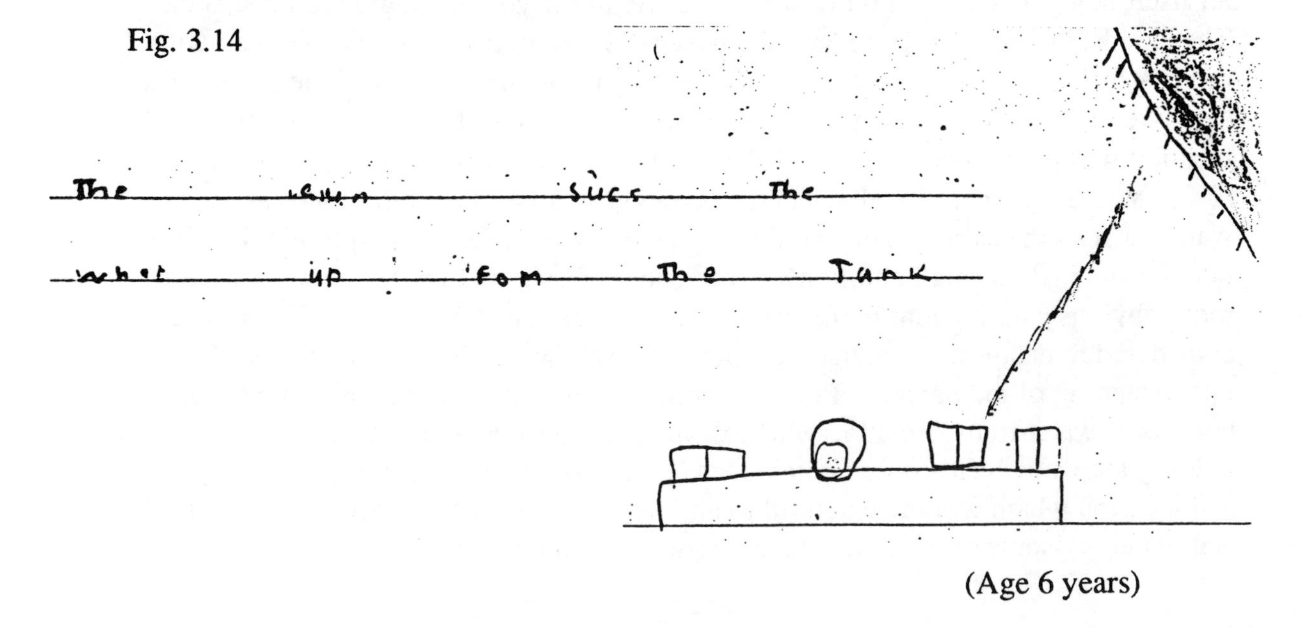

(Age 6 years)

"The sun sucks the water up from the tank."

b. Clothes drying

As previously mentioned, the responses to this activity can be grouped into the same categories as for the water tank. However, there are some interesting differences in frequencies of responses which might reflect the different nature of the activity and its links to familiar, everyday experiences.

The most prevalent response from infants was that the water dried up (33%), and this explanation had virtually disappeared by the upper juniors (3%). Nearly as common (30%) and not tailing off as much (13% in upper juniors) was the idea that the water went into the material (Fig. 3.15), or into another object.

Fig. 3.15

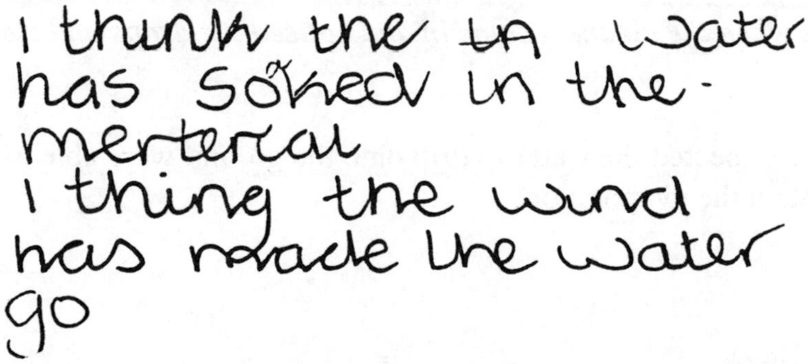

I think the water
has soked in the-
merterial
I thing the wind
has made the water
go

(Age 10 years)

> *"I think the water has soaked in the material. I think the wind has made the water go."*

Fig. 3.15 also illustrates how children appeared to be quite happy to give two different responses to a situation without feeling that one excluded the other.

One response which did not have an equivalent with regard to the water tank was that of the water from the washing dripping onto the floor. This was an accurate observation of what some of the water was likely to do, and the number of children mentioning it did not vary much with age. However, more younger than older children offered this as their complete explanation, sometimes suggesting the cleaners mopped it up.

Fig. 3.16

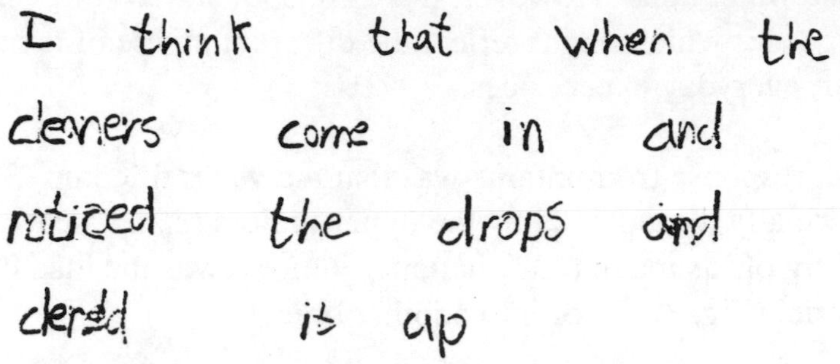

I think that when the
cleners come in and
noticed the drops and
clerdd it up

(Age 6 years)

"I think that when the cleaners came in and noticed the drops and cleaned it up."

Some children who expected the water to drip onto the ground were able to suggest a further step to explain the evaporation:

Fig. 3.17

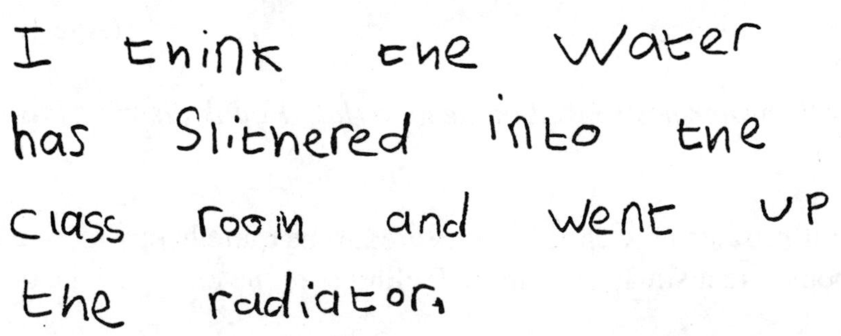

I think the water
has Slithered into the
class room and went up
the radiator.

(Age 6 years)

"I think the water has slithered into the class room and went up the radiator."

Fig. 3.18

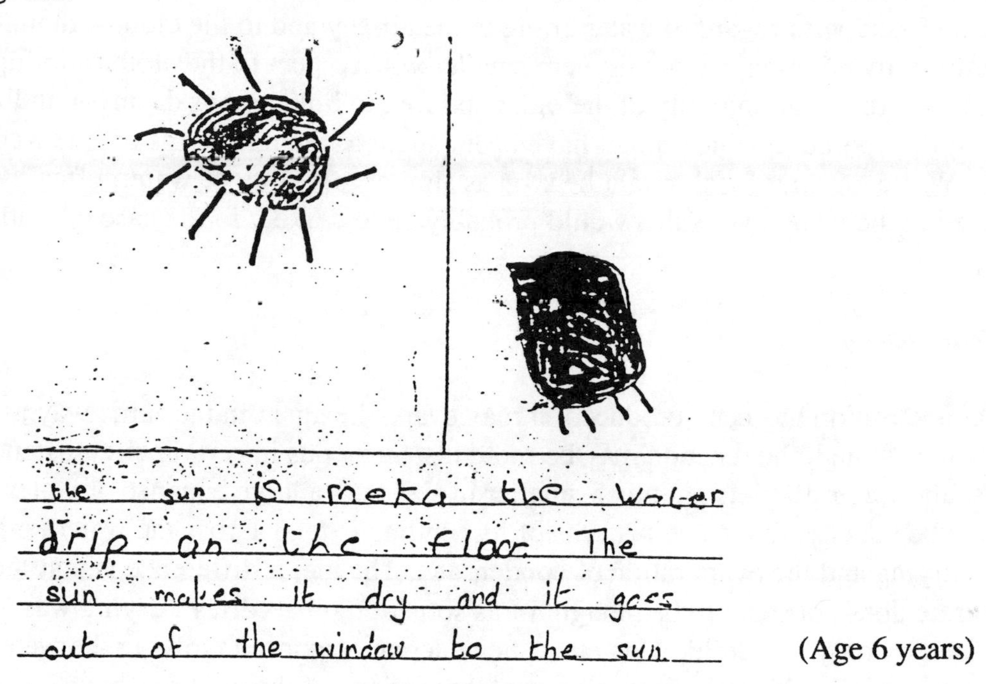

(Age 6 years)

"The sun is making the water drip on the floor. The sun makes it dry and it goes out of the window to the sun."

Fig. 3.19

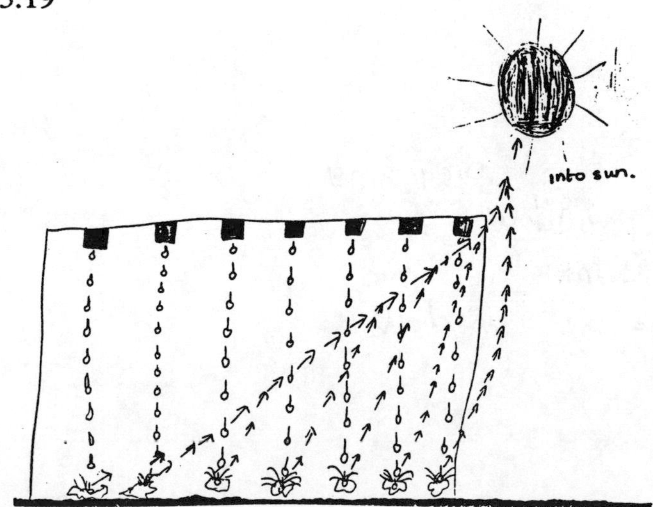

(Age 8 years)

The above examples illustrate the same causal relationship between the agent of evaporation, the sun, and the location to which the water moves, which was postulated in connection with the water tank. In fact, the same pattern of responding is found, with the peak at lower junior age (15%) and no respondents at upper junior age.

Very similar patterns of response to the water tank and the clothes drying activities were also found with regard to water going to the air/sky and to the clouds, though the numbers mentioning the clouds were smaller with respect to the clothes drying. This might indicate an inability of the older children to link clothes drying with the process of evaporation which might have been taught to them in terms of 'the water cycle'. While the water tank might be seen as similar to a body of water, often portrayed in the water cycle, this would probably be less likely in the case of clothes drying.

c. Condensation

Once condensation has been produced, it may evaporate again in the same way as other water, though the reaction may be more instantaneous in certain circumstances because the water droplets are much smaller in size than a large expanse of water. It seems to be the case that some patterns of responding extend across the water tank, clothes drying and the evaporation of condensate. The major difference is that the condensate does not seem to be thought of as something that 'dries'; drying was mentioned by only one child. However, the concrete response involving the water going into a surface is reminiscent of some descriptions of the clothes drying, with one-third of infants stating that the breath had gone into the window.

Fig 3.20

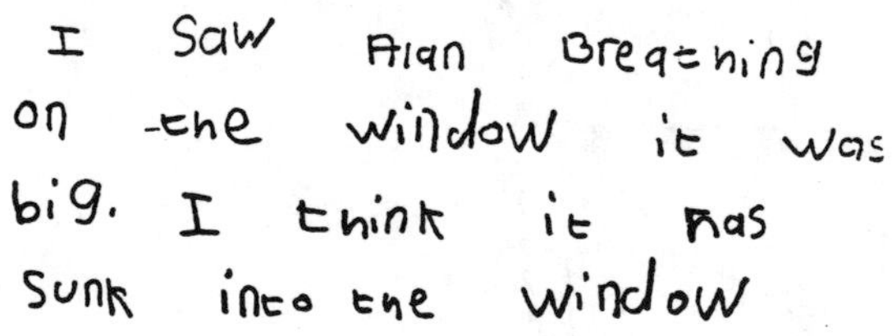

I Saw Alan Breathing
on the window it was
big. I think it has
Sunk into the window

(Age 6 years)

"I saw Alan breathing on the window it was big. I think it has sunk into the window."

Fig. 3.21

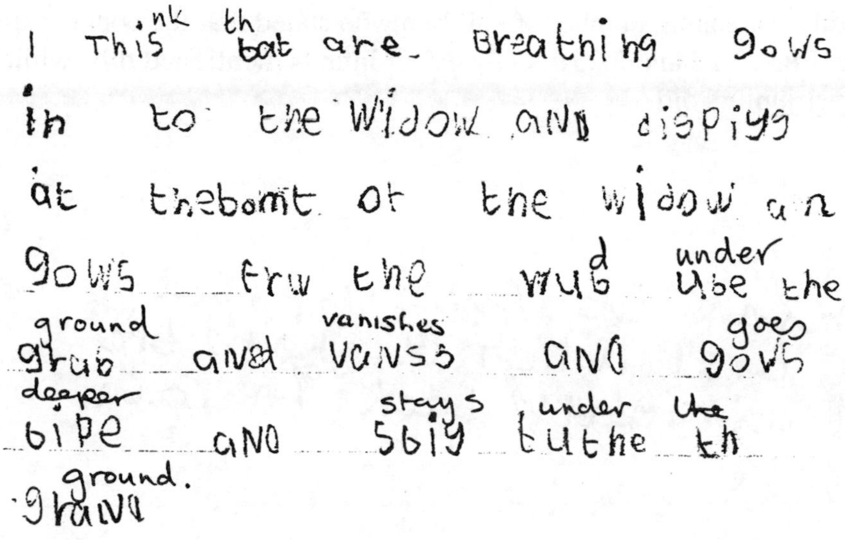

I This(ink) (th)bat are Breathing gows
in to the Widow and dispiys
at thebomt of the widow an
gows frw the wu(d)b (under) ube the
(ground) grub and (vanishes) vanss and (goes) govs
(deeper) bipe and (stays) sbiy (under) tuthe (the) th
(ground.) grawa

(Age 6 years)

"I think that our breathing goes in to the window and disappears at the back of the window and goes through the wood under the ground and vanishes and goes deeper and stays under the ground."

It might be worth adding a note of caution here since one child elaborated his response to give further insight into his meaning (Fig. 3.22). The word 'soaked' could seem to indicate that the water was sinking into the window. However, an alternative use of the word is when someone is 'soaking wet', implying that they are very soggy. The child could just mean that the window became very wet before the water went into the air.

Fig. 3.22

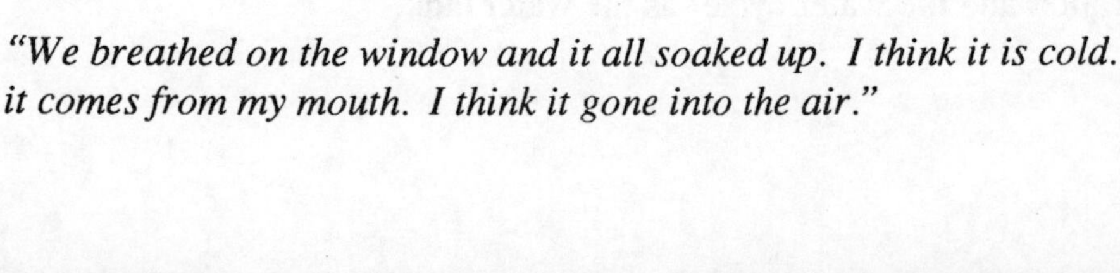

we breathed
on the window and it all somked up
I think it is cold I think it comes from
my Mouth I think it gone in to the
air

(Age 9 years)

"We breathed on the window and it all soaked up. I think it is cold. I think it comes from my mouth. I think it gone into the air."

Despite very little 'drying' being mentioned, there was a suggestion that water went to the sun. While the proportion of children suggesting this was very small (7%) they were lower juniors, which follows the same pattern as the other activities. Similarly, there was a steadily increasing number of children who stated that the condensation went into the air (Figs. 3.22 and 3.23). Only 4% of infants mentioned this, while over half the upper juniors did.

Fig. 3.23

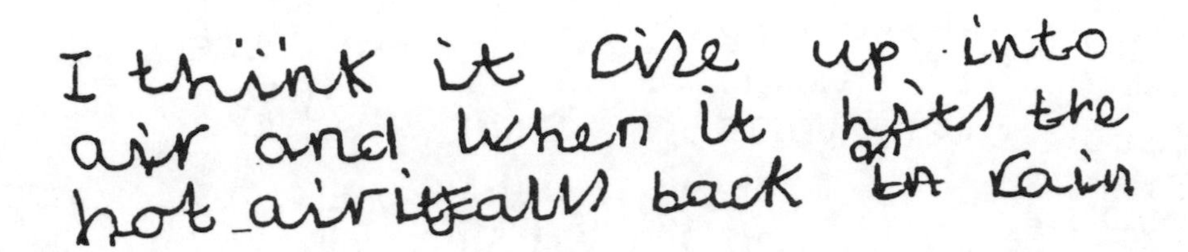

(Age 10 years)

"I think it rise up into air and when it hits the hot air it falls back as rain."

A finding which could support the possibility that the 'disappearance' of water mentioned by older children in connection with the water tank might be an inability to articulate notions of transformation is that few upper juniors talked about condensation disappearing. The rate of evaporation from a water tank is so slow that it could not easily be appreciated by young children. The evaporation of condensed water vapour, on the other hand, was readily observable and 'disappearance' was therefore a response which would be equally accessible to all ages, regardless of their level of development. Since fewer older children used the 'disappearance' explanation than in connection with the water tank, this could suggest that the water tank response may not represent an inability to conserve volume (as demonstrated by infants with the water tank and condensation) but a difficulty in describing a more complex phenomenon.

A few of the older children mentioned the condensation going to the clouds, or into rain though this number was small compared with the water tank. This could be, as with the clothes drying, that this source of evaporating water was not as clearly linked with 'evaporation and the water cycle' as the water tank.

What made the water go?

This question elicited both a wide range of necessary conditions for, and some interesting attempts at conveying agents and mechanisms of, evaporation.

The suggested causative agents of evaporation mirror quite closely the suggested locations for the evaporated water and some interesting comparisons can be made between the different activities.

a. Water tank

A substantial number of younger children, mainly infants, thought that the water had been removed by a human or animal agent and had either been drunk or tipped away somewhere (Fig. 3.24). This appears to represent attempts to explain an unobservable phenomenon in terms of an event which would have been observable, had they been there.

Fig. 3.24

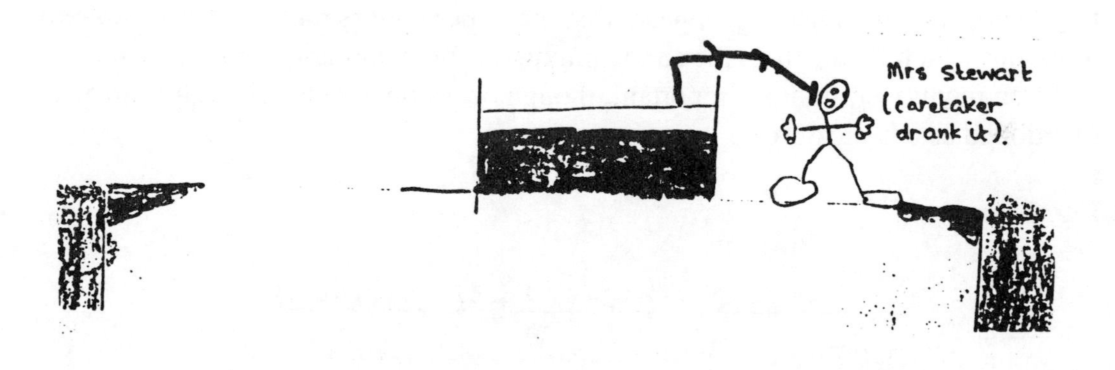

(Age 8 years)

"Mrs. Stewart (caretaker) drank it." [Teacher note]

This concrete notion was not suggested by upper juniors who may be able, to some extent, to comprehend very gradual, imperceptible change. One infant (Fig.3.25) stated clearly that the phenomenon happened in his absence.

Fig. 3.25

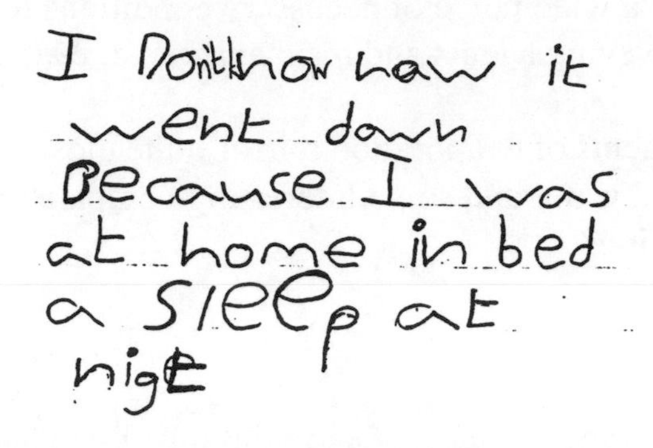

(Age 6 years)

"I don't know how it went down because I was at home in bed asleep at night."

Nearly half of the infants and lower juniors mentioned the sun to be responsible for the water going. This fell dramatically to one-fifth of upper juniors, while the reverse trend happened with heat. It appears that the upper juniors may have been able to work out which factor from the sun is important for evaporation. Virtually no children mentioned light as important, though occasionally electric lights are mentioned as a heat source.

Fig. 3.26

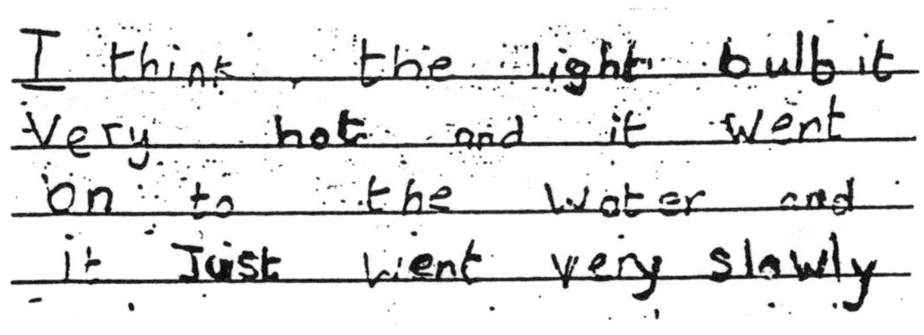

I think the light bulb it
Very hot and it went
on to the Water and
it Just went very slowly

(Age 8 years)

"I think the light-bulb it very hot and it went on to the water and it just went very slowly."

There was very little mention of the air or the wind as a contributory factor to evaporation from a water tank. This may be because the water tanks were indoors, and wind and air could be thought to be out of doors.

A number of different mechanisms was suggested by which the sun, and more occasionally air, clouds or heat, was thought to remove the water. These mechanisms seem to draw heavily upon analogies, particularly of magnets and drinking straws.

Fig. 3.27

(Age 6 years)

"The water goes straight up to the clouds."

Fig. 3.28

(Age 7 years)

"The sun is hot and the water is cold and the water sticks to the sun and then it goes down"

Fig. 3.29

Fig. 3.29

When the water ~~gose~~ evapairates it gose to a
cloud and then the cloud gose to
any place and lates it go outas rain
it will keepgoing till it is all gone
and then it will go to another ploce
with water and Do the same.
The cloud is like an magnite so the
water gose through
the craek's and gose up. that's what
I think

(Age 10 years)

"When the water evaporates it goes on a cloud and then the cloud goes in any place and later it go out as rain. It will keep going until it is all gone and then it will go to another place with water and do the same. The cloud is like a magnet so the water goes through the cracks and goes up, that is what I think."

Fig. 3.30

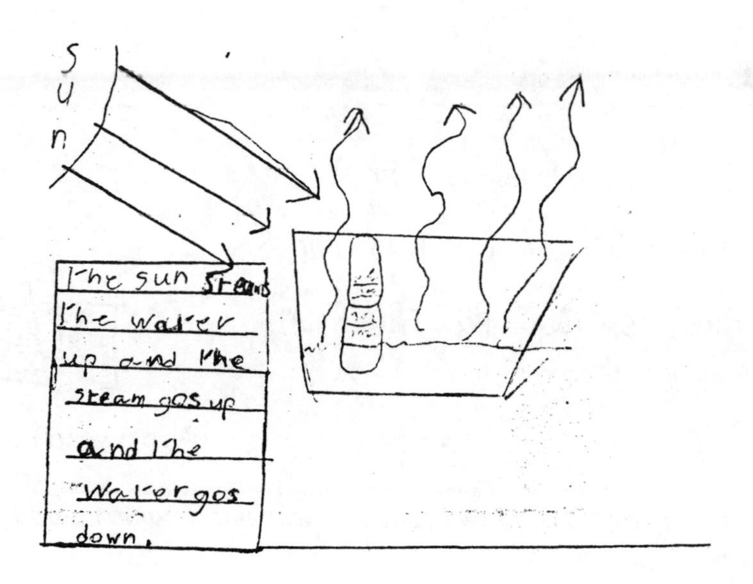

(Age 8 years)

"The sun steams the water up and the steam goes up and the water goes down."

Fig. 3.31

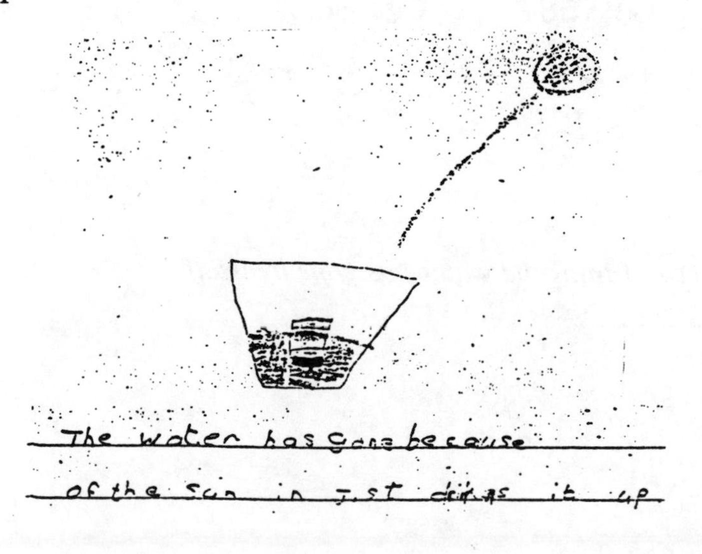

(Age 6 years)

"The water has gone because of the sun it just dries it up."

Infant children whose ideas suggested that the water moves to a nearby location could at the same time suggest that heat was important rather than the sun. While this could appear to be a more mature response, in a school environment, heat may be very much in evidence.

Fig. 3.32

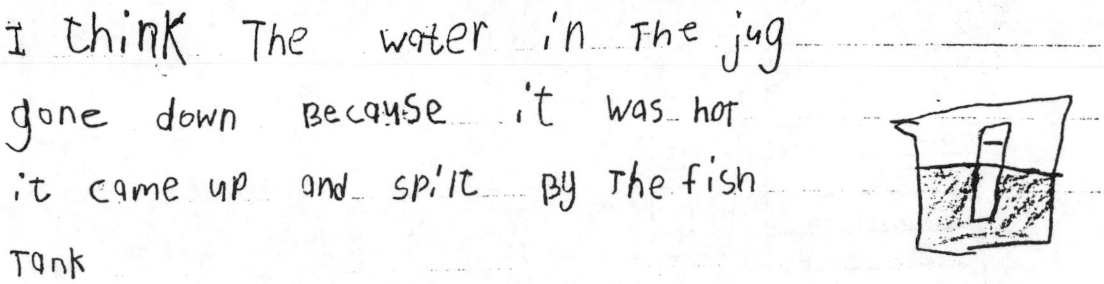

I think The water in The jug
gone down Becayse it was hot
it came up and spilt By The fish
Tank

(Age 6 years)

"I think the water in the jug gone down because it was hot it came up and spilt by the fish tank."

Three upper junior children suggested that the water moved itself with no obvious factor being involved:

Fig. 3.33

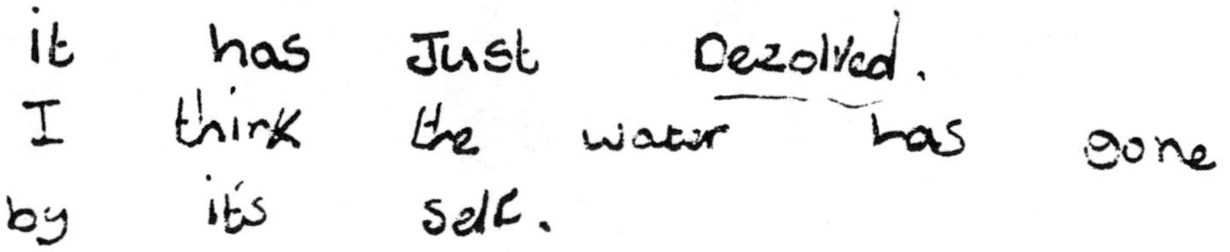

it has Just Dezolved.
I think the water has gone
by its self.

(Age 10 years)

"It has just dissolved. I think the water has gone by itself."

Fig. 3.34

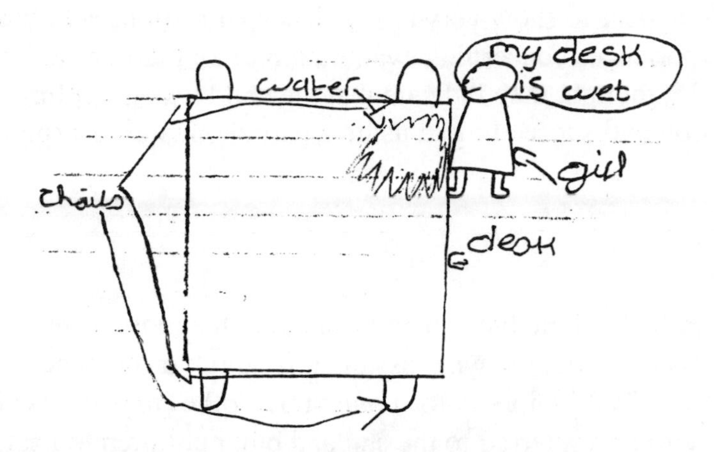

I think that the water will lift it self out of the box and land on one of the tables and when we all come back to school the next day Somebody will find that thier desk has got water on it.

(Age 10 years)

"I think that the water will lift itself out of the box and land on one of the tables and when we all come back to school the next day somebody will find that this desk has got water on it."

Fig. 3.35

I think the water has split up into millions of micro bits and floated up and ~~floating~~ it floats out of the doors or windows when they are opened.

(Age 10 years)

"I think the water has split up into millions of micro bits and floated up and it floats out of the doors or windows when they are opened."

Figures 3.33 and 3.34 appear to show very unsophisticated responses indicating either a complete lack of experiences concerning evaporation or an inability to link cause and effect. Figure 3.35, though, may indicate that the child was grappling with the notion of transformation and succeeding in using a particle model to explain the evaporation.

b. Clothes Drying

In contrast to the water tank where the sun and heat were the most commonly mentioned agents of evaporation, the wind assumed far more prominence in both the upper (35%) and lower (17%) juniors in the context of clothes drying. While with the water tank young children referred to the sun and older children to heat, this does not occur with this activity, there being more mention of sun than heat in each age group. These patterns of response suggest the existence of context effects which may be due in part to the clothes being dried, where possible, outside. Additionally, there were several mentions of tumble dryers, suggesting that childhood experiences are changing.

Again, as with the water tank, there were attempts to articulate actual mechanisms for the change. The range of mechanisms was very similar though with a higher occurrence of 'drying'. This might reflect the everyday nature of the activity since clothes are hung up 'to dry' and there is no need for further elaboration of the happening.

Fig. 3.36

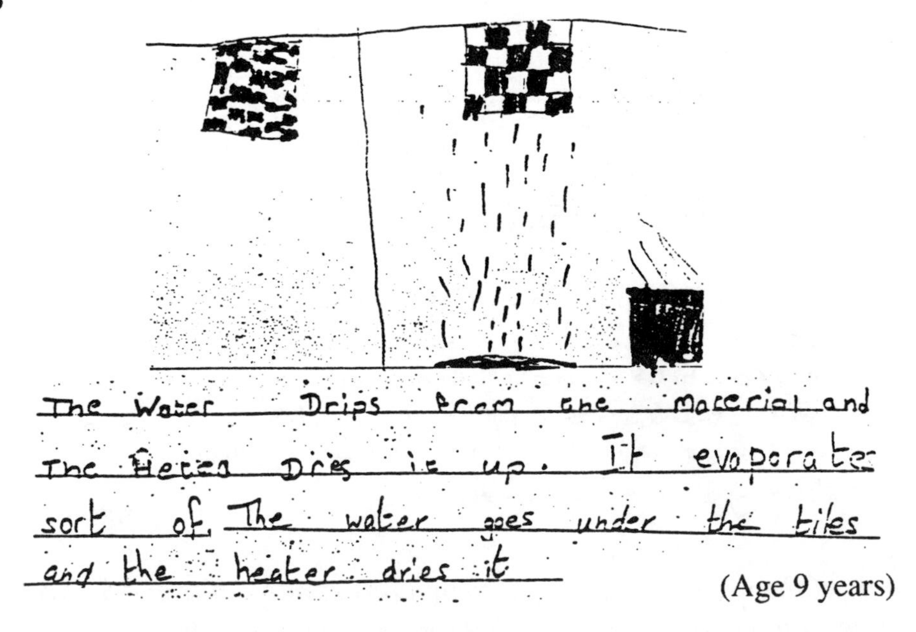

(Age 9 years)

"The water drips from the material and the heater dries it up. It evaporates sort of. The water was under the tiles and the heater dries it."

Fig. 3.37

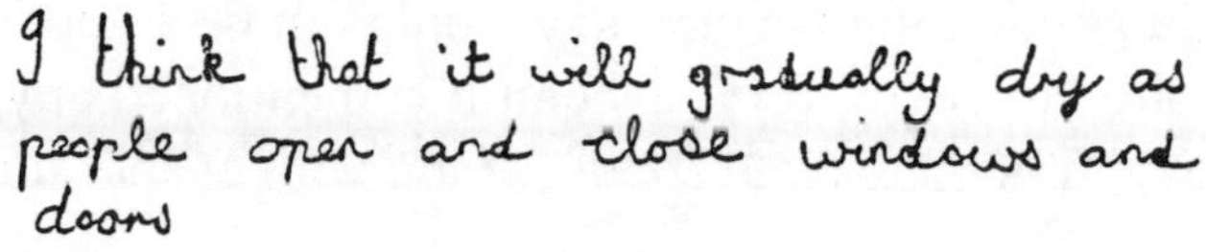

I think that it will grsdually dry as people open and close windows and doors

(Age 10 years)

"I think that it will gradually dry as people open and close windows and doors."

Fig. 3.38

(Age 9 years)

"The sun sucks up the water into the sun."

This suggests that condensation could be an important tool in the teaching of evaporation since it circumvents the sun and appears to encourage children to think about directly relevant factors. This tendency may also be related to prevalent weather conditions: a good clothes drying day would not be a good day for seeing condensation in the air. However, very few children mentioned any agent at all, possibly because they did not associate condensation with evaporation, or even with water.

Fig. 3.39

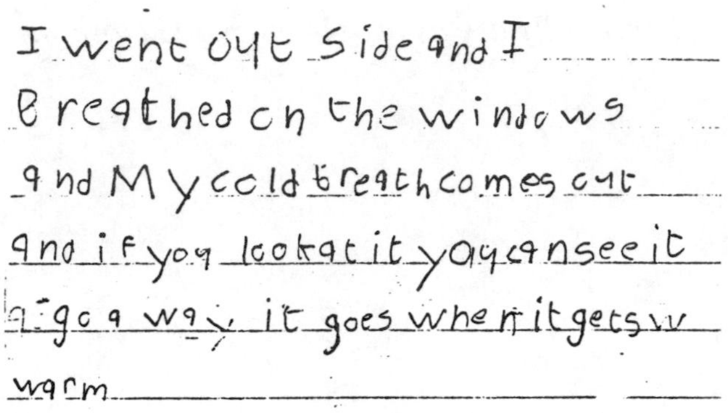

I went out side and I
Breathed on the windows
and My cold breath comes out
and if you look at it you can see it
a go a way it goes when it gets w
warm

(Age 6 years)

"I went outside and I breathed on the windows and my cold breath comes out and if you look at it you can see it go away it goes when it gets warm."

c. Condensation

The sun was not mentioned at all as a causative agent of evaporation, only heat and wind.

Can you alter the rate at which the water goes?

This question was asked of children in two parts: "Can you make the water go faster?" and "Can you make the water last longer?" It has also been analysed in that way, and some interesting differences emerge between children's ideas about speeding up and slowing down evaporation. Additionally, there are contrasts between upper and lower juniors.

Table 3.1 Can you make the water go faster/last longer? (Predominant responses are in upper case)

	Fish tank		Clothes drying	
	Can you make the water - go faster?	- last longer?	Can you make the water - go faster?	- last longer?
lower junior	CANNOT HEAT	NO HEAT NO SUN cannot	REMOVE WATER sun wind hot moving air cannot	ADD WATER NO HEAT NO SUN NO WIND cannot
upper junior	CANNOT HEAT remove water	CANNOT cover it no heat	WIND cannot hot moving air heat sun	DAMP PLACE cannot cover it no wind no heat

a. Fish tank

The most common response to the question concerning increasing the rate of evaporation was that this was something it was not possible to do. Of those children who considered it to be possible to make the water evaporate more quickly, the most commonly suggested method was the application of heat. One extreme example is shown in Fig. 3.40.

Fig. 3.40

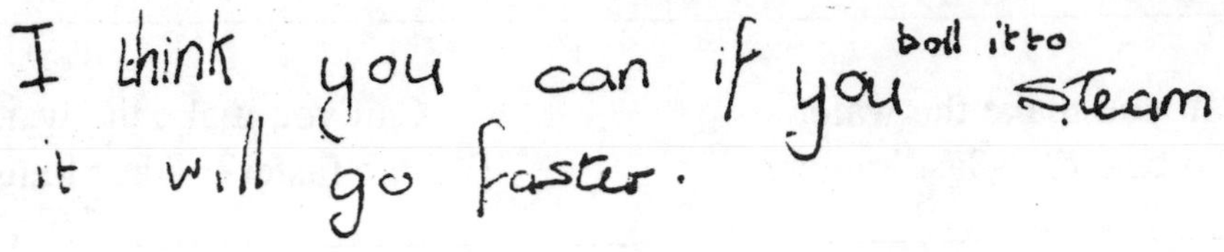

I think you can if you boil itto steam it will go faster.

(Age 10 years)

"I think you can if you boil it to steam it will go faster."

To make the water last longer, the lower juniors suggested keeping the tank away from the heat or sun, thus reversing the effect on evaporation by reversing conditions. (Fig. 3.41). Many upper juniors surprisingly responded that this rate alteration was also impossible. However, a group of older children suggested covering the tank (Fig. 3.42).

Fig. 3.41

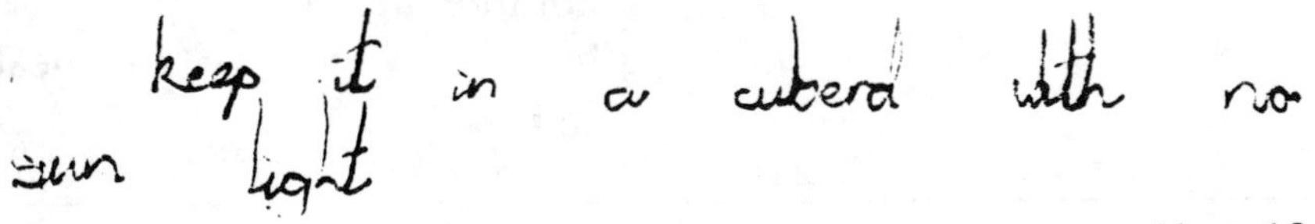

keep it in a cuberd with no sun light

(Age 10 years)

"Keep it in a cupboard with no sun light."

Fig. 3.42

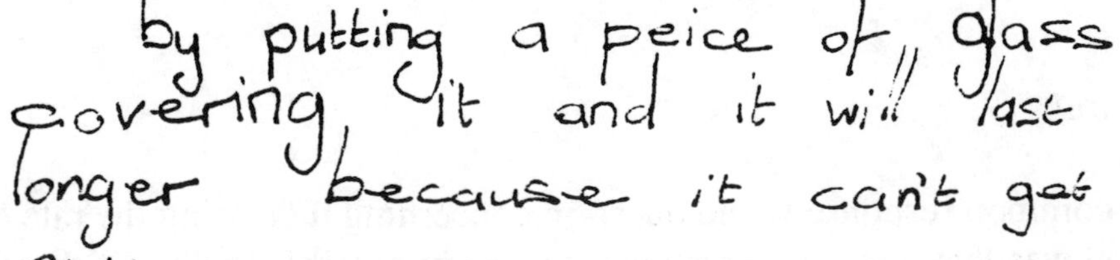

by putting a peice of glass covering it and it will last longer because it can't get out.

(Age 10 years)

"By putting a piece of glass covering it and it will last longer because it can't get out."

This explanation suggests that, even if covering was an intuitive response, based upon experience, some children were becoming more able to dissociate evaporation from a particular environmental factor. This ability might lead them towards a more general theory of evaporation than might the 'water cycle'.

A difference also appeared between numbers of children considering an increase or decrease in the rate of evaporation to be possible. More pupils felt that the water could be made to last longer than felt that it could be made to go faster.

Fig. 3.43

(Age 7 years)

> *"I think the water has dried up with the sun shining on it. I do not think the water can go fast but the water can go slow.**
> **Close blinds* ***and*** *stop the sun coming in to dry it up." (Teacher note)*

This difference in the ease with which children could suggest alterations in the rate of evaporation could possibly be related to a lack of comprehension of all the processes involved. This might make it easier to suggest ways of trying to retain the present water level, by slowing down water loss, than to find ways of making the change faster.

(b) Clothes drying

A much smaller proportion of children than in connection with the water tank felt that the rate of evaporation could not be changed; perhaps this was because drying clothes is a more familiar activity than a tank of water. However, in the same way as for the water tank more upper than lower juniors felt that the rate could not be changed, and more felt it could not be accelerated than felt it could not be slowed.

Fig. 3.44

I think the heat as made the water go. the sun take it. time.

(Age 10 years)

"I think the heat has made the water go. The sun takes its time."

In order to make clothes dry faster, the most common response from lower juniors was to remove some water, for example by wringing the material out. This idea was not very common from the upper juniors. With regard to the water tank the pattern of suggesting the removal of water was reversed, with more upper juniors suggesting it. Other means of accelerating drying were thought by lower juniors to be the sun, the wind and hot moving air (e.g. a tumble dryer). Upper juniors responded similarly except that wind was thought to be more significant, and heat was also included.

To slow down the rate of evaporation, lower juniors thought it would be important to keep the wet material away from heat, the sun and the wind, again a reversal of conditions to reverse the effect. Upper juniors, in a similar way to their responses to the water tank, felt that the wet material should be put in a damp place, covered, or kept away from heat or wind. This selection of responses suggests that a more sophisticated understanding might be developing amongst some upper junior children which enables them to make use of the process of evaporation in constructing a response to the question.

The responses of children to these two questions about each of two activities enable some tentative explanations to be offered. In response to both activities, lower junior children (who responded positively) gave answers which appeared logical: they reversed the prevailing condition in order to reverse the effect on evaporation; thus if heat would speed up the process then a lack of heat would slow it down. This pattern of responding did not occur with upper junior children, who seemed to be more likely to reject the idea of affecting the process. This negative response was particularly marked in connection with the water tank and it is possible to speculate that the older children were more likely to have learnt about the water cycle. This teaching may not have been extended to include alterations of rate and as the water tank resembles the sort of water source often included in water cycle diagrams, e.g. a lake, the teaching might not have encouraged the children to relate their everyday experiences to what they had learnt.

The learning experiences associated with the water cycle may have been presented in such a way as to suggest that all water 'evaporates' from the sea or other large expanses of water. The water tank, which contains a large amount of water, might be close enough to this taught model for the teaching to impair children's use of their everyday experience in this context. Thus, while younger children were able to suggest ways of altering the rate of evaporation in a logical manner, some older children appear either not to have been able to, or to have used concrete explanations such as removing water from the water tank to make the water go faster. The clothes drying, on the other hand, might be far enough removed from the standard model that children's intuitive thinking is more evident and they seem to use their observations to help them explain the process.

The ideas of some upper juniors, however, appear to have developed in such a way that the children were able to suggest conditions for altering the rate of evaporation which were not dependent upon the causative agents, e.g. covering the water or putting it in a damp place. This dissociation from the observable agents of change suggests that the children have made observations which might allow their thinking to develop to encompass a much broader notion of evaporation than could be implied by the water cycle.

There are examples of children who appeared to have a good knowledge of the 'water cycle' but when questioned about clothes drying showed that they were thinking at a concrete level. This suggests that they may not be applying their acquired knowledge across a range of difference contexts.

Fig. 3.45

Today we put some water into a fish tank and into a saucer. I think that the water will turn into vapor. And to follow the rain cyclcle.

Washing. I think that the water will run to the bottom and will drop on the floor untill it dry's

(Age 10 years)

> *"Today we put some water into a fish tank and into a saucer. I think that the water will turn into vapour. And to follow the rain cycle.*
> *Washing. I think that the water will run to the bottom and will drop on the floor until it dries."*

Can you get the water back?

There appears to be very little idea that the water which evaporated could be retrieved back. A few older children gave reasons for their belief.

Fig. 3.46

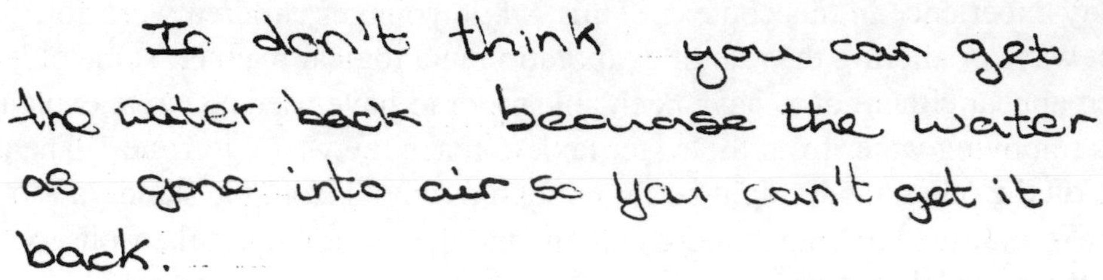

I don't think you can get
the water back because the water
as gone into air so you can't get it
back.

(Age 10 years)

"I don't think you can get the water back because the water has gone into the air so you can't get it back."

Fig. 3.47

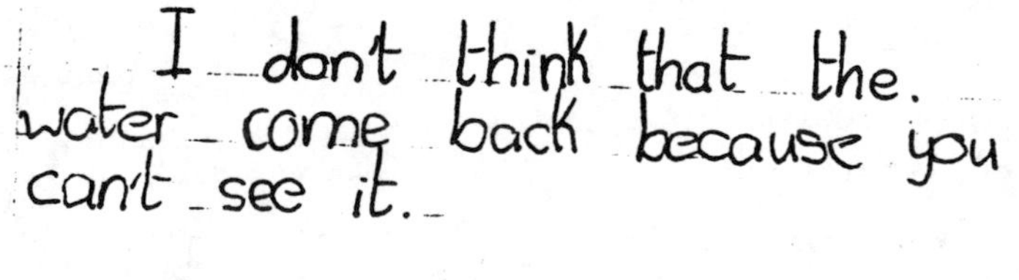

I dont think that the
water come back because you
can't see it.

(Age 10 years)

"I don't think that the water come back because you can't see it."

This question was revealing since it clarified that certain children who thought they 'knew' the answers had ideas which were inconsistent with each other; Fig. 3.46 is by a boy who also wrote about water vapour and steam. Some children who did feel that it was possible to get the water back also revealed a lack of understanding of evaporation.

Fig. 3.48

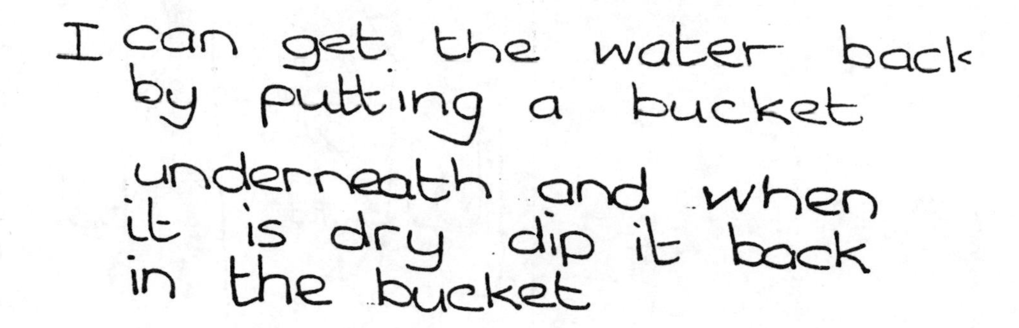

(Age 10 years)

"I can get the water back by putting a bucket underneath and when it is dry dip it back in the bucket."

Of those children who thought that the water could be retrieved as rain, more of them were responding with regard to the water tank than the clothes drying, and also at a younger age. This is consistent with the manner in which children responded to questions about where the water had gone, and also about the rate of evaporation, and again might relate to previous teaching of the water cycle, which emphasises a large body of water as the source of rain.

What is condensation?

In order not to lead children to refer to water in the context of condensation, questions were worded in a way that did not use the word. Children were asked what they thought 'it' was. The most common response from infants and lower juniors was that it was 'breath', though lower juniors also suggested 'air from my body' and 'steam' with a reasonable frequency. Upper juniors considered it to be one of steam/fog/mist, while 'air from my body' was still mentioned to some extent. A small proportion of children of each age described it as being 'smoke'.

Fig. 3.49

I went out Side and Started breathing out of my mouth. Somthing came out wich was just about visible and then it went invisible into the air. I think it was my breath gone frozen because when it is a warm day and you breath out of your mouth nothing comes out.

(Age 10 years)

"I went outside and started breathing out of my mouth. Something came out which was just about visible and then it went invisible into the air. I think it was my breath gone frozen because when it is a warm day and you breathe out of your mouth nothing comes out."

Fig. 3.50

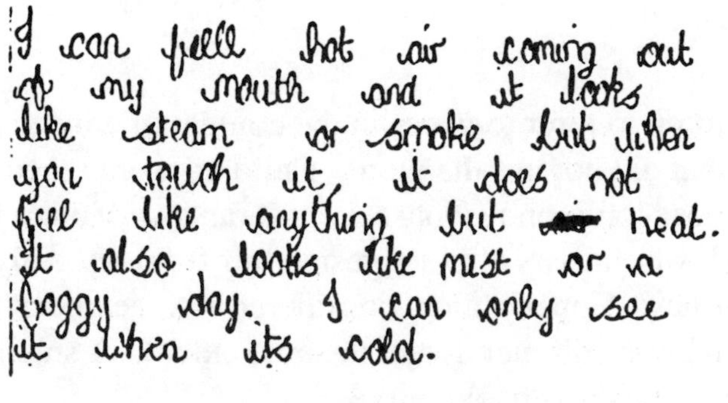

I can feell hot air coming out of my mouth and it looks like steam or smoke but when you touch it, it does not feel like anything but heat. It also looks like mist or a foggy day. I can only see it when its cold.

(Age 10 years)

"I can feel hot air coming out of my mouth and it looks like steam or smoke but when you touch it, it does not feel like anything but heat. It also looks like mist or a foggy day. I can only see it when it's cold."

Fig. 3.51

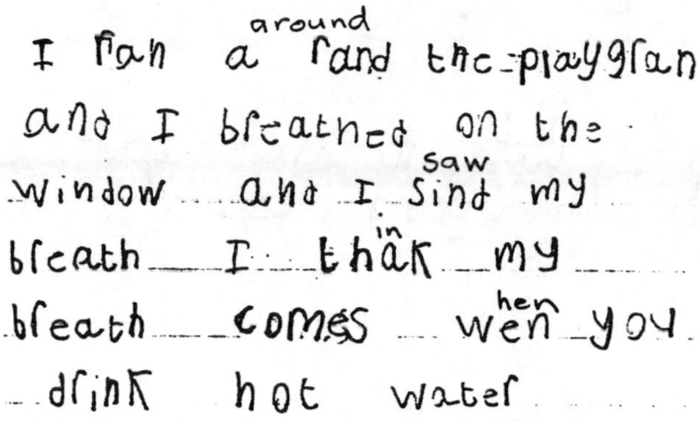

I ran a [around] rand the playgran
and I breatned on the
window and I sind [saw] my
breath I thäk [in] my
breath comes wen [hen] you
drink hot water

(Age 6 years)

"I ran around the playground and I breathed on the window and I saw my breath. I think my breath comes when you drink hot water."

There was very little mention of water being present. Those children who did refer to water were amongst the upper juniors.

Fig. 3.52

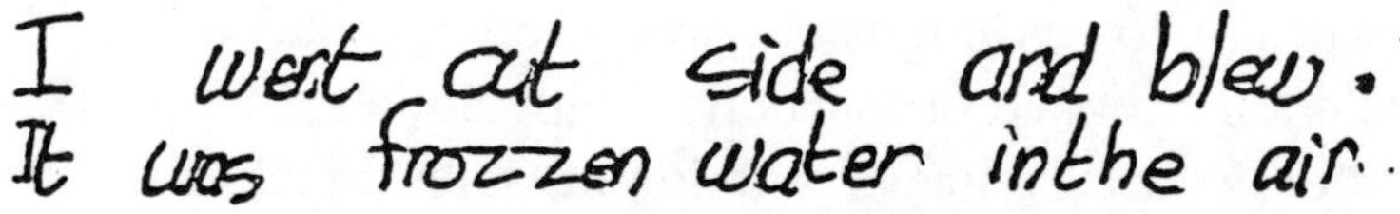

I went at side and blew.
It was frozen water inthe air.

(Age 10 years)

"I went outside and blew. It was frozen water in the air."

Fig. 3.53

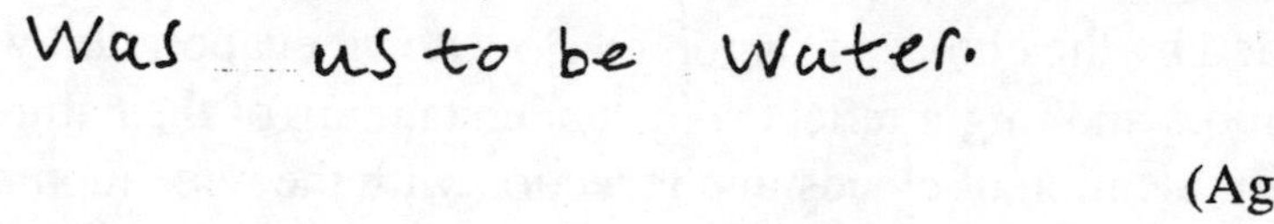

I think it is steam What
Was us to be Water.

(Age 10 years)

"I think it is steam what was used to be water."

What has made the condensation appear?

This question was responded to with greater frequency by upper juniors. Very few infants addressed it directly. It was quite often stated that temperature was an important factor, either the temperature of the breath, or of the air/window, sometimes the combination of hot breath and cold air meeting (Fig. 3.54).

Fig. 3.54

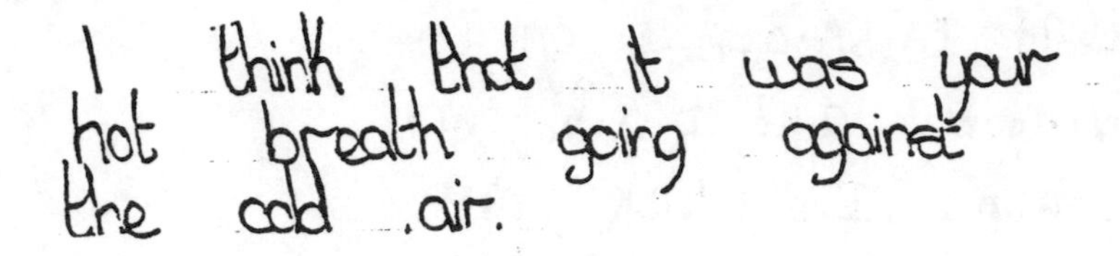

(Age 10 years)

"I think that it was your hot breath going against the cold air."

The explanation offered in Fig. 3.54 was worded in such a way that it seems likely that this type of definition was presented to the child in a formal learning situation; it is unlikely that this child would have discovered it in this form for themselves.

Use of the word 'Evaporation'

The word 'evaporation' was used by over half of the children in the upper junior sample in connection with both the water tank and the clothes drying. Interestingly, the use of the word by lower juniors only occurred in connection with the water tank (20%) and was not mentioned in connection with clothes drying. This pattern of usage is similar to the occurrence of a notion of water transformation (though with much higher frequencies) and also the idea that the water can reappear as rain.

It is possible that the idea of evaporation is conceptually easier to grasp in connection with an obvious quantity of pure water. Alternatively, children might have been exposed to the idea of the water cycle which is usually portrayed as water evaporating from a vast expanse. This suggests that children of lower junior age may be able to apply acquired knowledge in very limited contexts, those which present a perceptually similar experience to that used in the teaching situation. There may be very little understanding of the process behind the cycle until children have reached the age of ten or eleven, and even then, only by a limited number of children. This position is supported by the choice of factors to slow down evaporation which was made by upper juniors and might reflect some understanding of the nature of the process. Consistent mention of clouds in connection with the relocation of the water also follows this pattern and may also be associated with formal teaching of the water cycle, since younger children only hold this idea in connection with the water tank.

Fig. 3.55

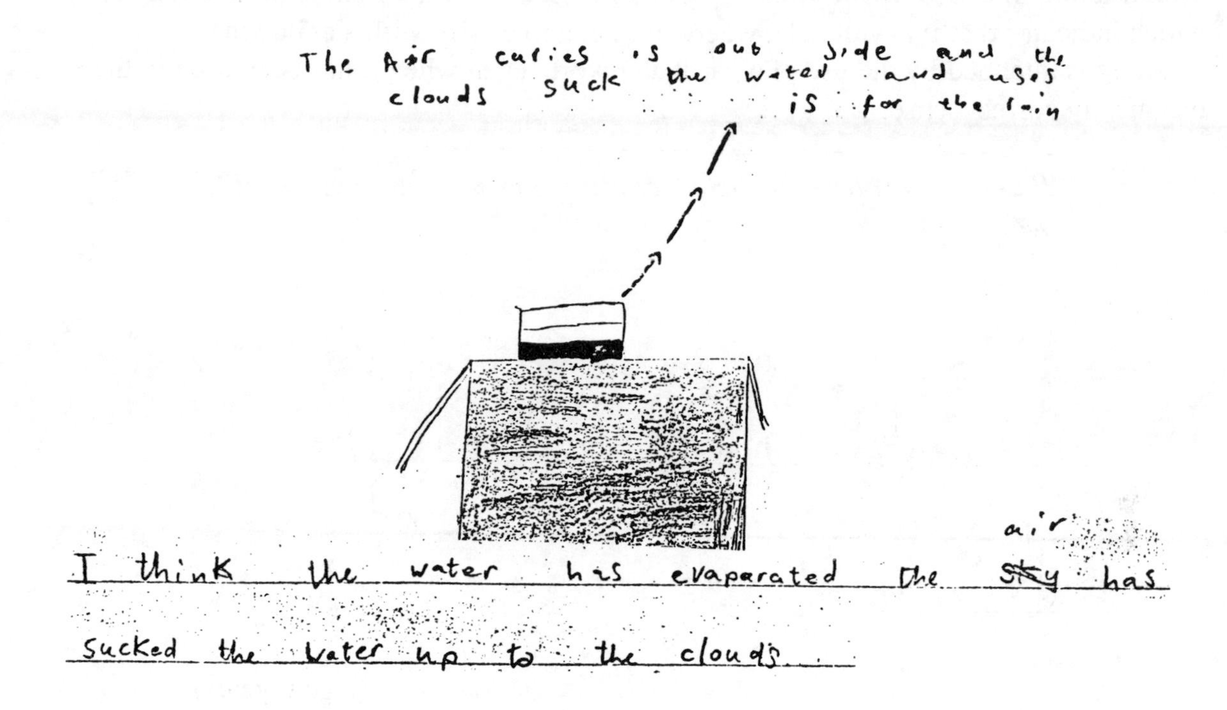

(Age 8 years)

"The air carries it outside and the clouds such the water and uses it for the rain. I think the water has evaporated the air has sucked the water up to the clouds."

Fig. 3.56

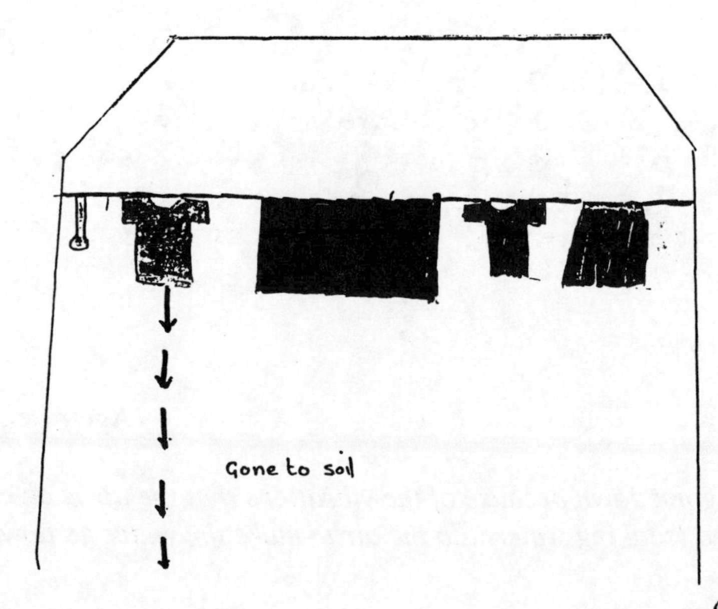

(Age 8 years)

The meaning intended by the word 'evaporated' seems to vary from child to child which indicates that it is valuable, where possible, to clarify with a child what meaning is intended by the use of a particular word, even when it has been used in the technically correct sense.

> *e.g. "Water evaporates onto the window as condensation. Evaporate means to go up."*

Fig. 3.57

evaporeated meens that it goes up into the air and dissappeared

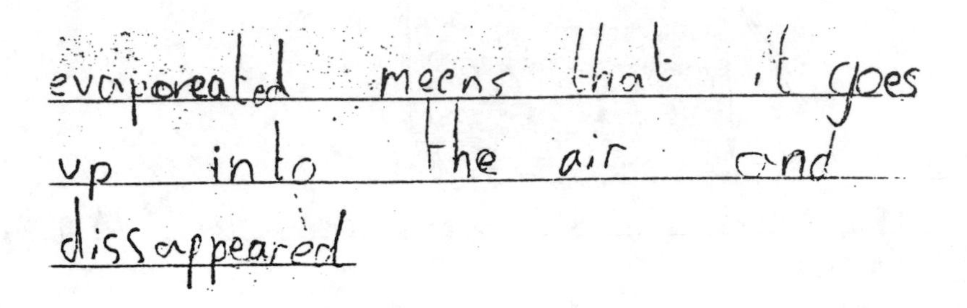

(Age 6 years)

"Evaporated means that it goes up into the air and disappeared."

There are also examples of children confusing the word 'vibration' with 'evaporation' since they sound very similar.

Fig. 3.58

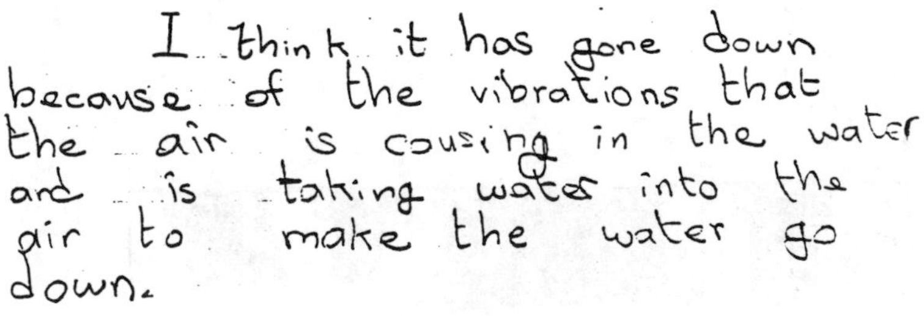

I think it has gone down because of the vibrations that the air is cousing in the water and is taking water into the air to make the water go down.

(Age 8 years)

"I think it has gone down because of the vibrations that the air is causing in the water and is taking water into the air to make the water go down."

The converse happens in discussion concerning sound vibrations where evaporation is mistakenly used. (There is no possibility that the confusion emanates from molecular motion in these instances).

Use of the word 'Condensation'

This word was used extremely rarely and never in connection with the activity intended to create condensation. The everyday definition of condensation seems to be concerned with water on the inside of windows and little, if any, connection appears to be made with the process of evaporation, of which it is the converse.

Fig. 3.59 *"It evaporates onto the window as condensation. When it's cold it sticks to the window and you can draw on it."*

4. INTERVENTION

The Exploration phase enabled children to experience the phenomena of evaporation and condensation within the classroom. During this period teachers had endeavoured to enable children to express their thoughts and ideas in an open, non-directive manner. Many teachers had been surprised, and encouraged, firstly at the existence of the children's ideas and secondly at the way they had been able to obtain access to them by using the elicitation techniques.

The response of both children and teachers to the elicitation work led the research team to develop some general strategies for Intervention which the teachers could use in their classrooms with their children.

The four strategies were designed to help teachers structure activities in four areas, each of which appeared to be capable of exerting important influences in the formation of children's ideas.

a. Helping children to test their own ideas

Children's ideas were often based upon experiences including direct observation, or upon situations which might have been presented to them as 'something to learn'. In either case the ideas often seemed to be context-specific, formulated on the basis of limited experience in which available information could have been incomplete or misinterpreted.

Ideas thus formed were often at odds with carefully made observations. It was envisaged that, by encouraging children to test their ideas in a more considered and rigorous way, i.e. in a scientific manner, they might be enabled to develop their thinking along more productive lines.

Fig. 4.1

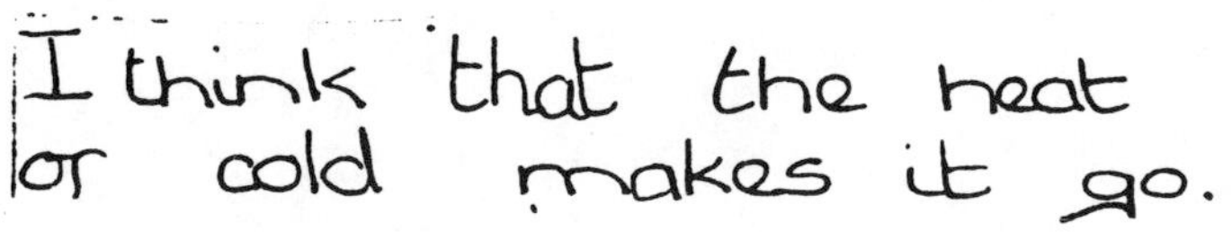

I think that the heat
or cold makes it go.

(Age 10 years)

"I think that the heat or cold makes it go."

The example in Fig. 4.1 shows an idea containing two alternatives. A carefully controlled test involving equal amounts of water in locations at different temperatures might enable this child to develop her understanding of a particular idea in a way which would make it more accurate scientifically and more useful as a way of explaining evaporation; a distinction between a cold air temperature and a cold wind would remove the contradiction implicit in the quoted statement.

It was felt by the research team that, in order for the investigation to engage as fully as possible with the child's thinking, it was important for the children themselves to suggest the tests. The teacher's role, rather than being instructional, was to ensure that the children carried out their experiments in a fair way, using equal volumes of water, making accurate measurements and following their own plans. This role required teachers to develop a questioning technique which would encourage children to pursue the implications of their ideas in action in a disciplined manner. This strategy, of children testing their ideas, was to be the main focus of each class's Intervention experiences.

b. Encouraging children to generalise from one specific context to other through discussion

It has been mentioned during the discussion of children's ideas in Section 3 that many ideas appear to be context-specific and that children seem to see the different activities as unrelated.

Fig. 4.2

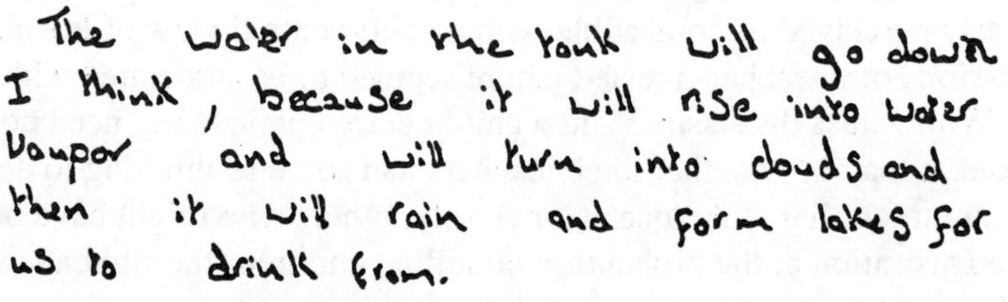

The water in the tank will go down
I think, because it will rise into water
Vaupor and will turn into clouds and
then it will rain and form lakes for
us to drink from.

"The water in the tank will go down I think, because it will rise into water vapour and will turn into clouds and then it will rain and form lakes for us to drink from."

Fig. 4.3

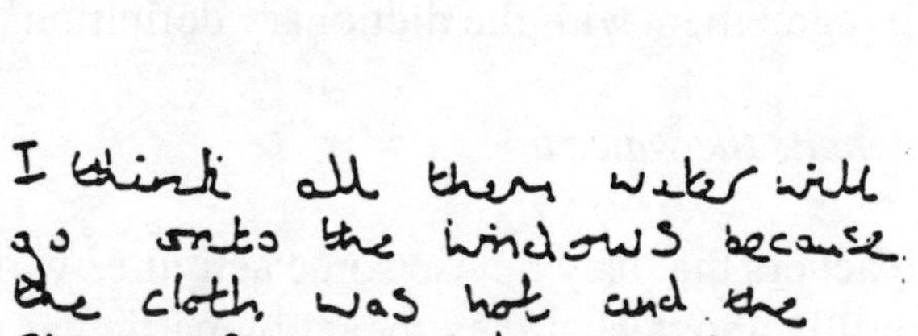

I think all them water will
go onto the windows because
the cloth was hot and the
steam of the cloth will go
onto the windows.

"I think all the water will go onto the windows because the cloth was hot and the steam off the cloth will go onto the windows."

Fig. 4.4

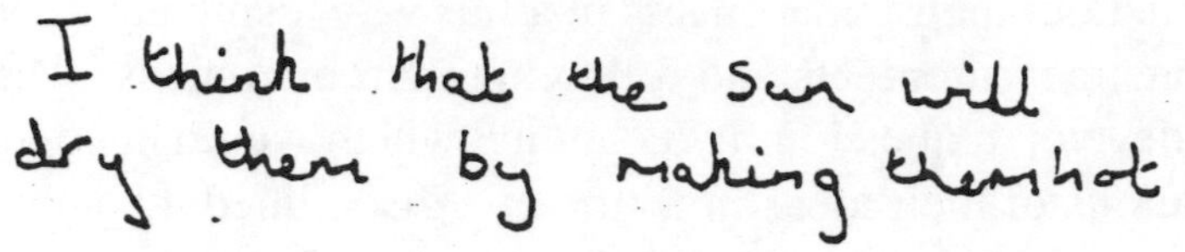

I think that the sun will
dry them by making them hot

(Age 10 years)

"I think that the sun will dry them by making them hot."

It was proposed that teachers could help children to see the similarities between the different contexts and thus help them to develop their ideas in a way which would make them applicable to more than one situation and thus more useful in explaining the phenomenon of evaporation. Once again, the management technique is an extremely subtle one which should stop short of the teacher making the connections for the children and handing them over. The skill was to set up discussions which would enhance the possibility of children making the links and generalisations for themselves. It was interesting to hear from the teachers that this technique was not always clearly perceived as compatible with a child-centred view of learning. Whole class discussions of a teacher-directed nature seemed to be associated with a didactic approach. While class discussions and a child-centred philosophy need not be incompatible, it may be true that some teachers had some re-thinking to do to come to terms with an unfamiliar technique. Conversely, others who might have been building up frustration at the prohibition on telling children 'the right answer' might have faced struggles of a different kind if their habitual style were of a more didactic nature.

c. *Encouraging children to develop more specific definitions for the words they used*

It became clear that many children used words in a manner which was idiosyncratic and not necessarily in agreement with the dictionary definition. For example,

"The sun melts the water up".

It was suggested to teachers that they devise some activities which would encourage children to think carefully about what they meant by particular words. The emphasis was on activity, so that the context in which the word was being used was explicit. It was anticipated that this approach might enable children to come to a consensus definition for particular words, especially everyday terms which children seem to use to describe the processes of evaporation and condensation, e.g. 'dry', 'dissolve', 'melt'. The emphasis was intended to be on children clarifying words they already used rather than the introduction of scientific vocabulary, though it was acknowledged that this might become appropriate if children were clearly defining a process for which they lacked a precise name.

d. Finding ways to make imperceptible changes perceptible

The process of evaporation involves liquid water changing into an invisible vapour and moving into the air. These changes are difficult to 'see' since the product is invisible, and are made more intangible since they often occur over a prolonged period of time. The need for young children to explain a very gradual change in terms which are accessible to them often makes them suggest that the change must happen in their absence, since they cannot see it. Conversely, condensation appears from 'thin air' and the instant appearance needs to be explained. It was suggested to teachers that they might try to find a way of making these imperceptible changes perceptible for the pupils.

These four strategies were presented to teachers for them to apply with their classes as appropriate. It was suggested that the emphasis of classroom work should be on children testing out their ideas and each teacher was asked to ensure that they also included one 'activity' related to generalising and one focusing on vocabulary. A minimum of four sessions of SPACE work over the four weeks of the Intervention was requested. The Intervention guidelines given to teachers can be found in Appendix VI.

The following examples illustrate the manner in which teachers implemented the Intervention strategies in their classrooms.

a. Helping children to test their ideas

i. The sun shining on the wet cloth made the water go.
(age 5-6 years)

A group of infant children felt that the wet cloth dried because the sun shone on it. To test this idea they decided to put a recently washed cloth into a cupboard which was covered with black paper to keep the light out. They predicted that the cloth in the cupboard would remain wet, not drying at all, because the sun could not shine on it. The following day, the cloth was examined and found to be dry. After initial surprise the children agreed with one child's assertion that it had dried because the cupboard was 'warm and cosy'. These children appeared to have considered the sun's effects to be localised to the particular places upon which it shines. Through this investigation they had found that direct sunlight was not the main cause of water going, and they had also raised an alternative, that heat might be important. This possibility formed the basis of their next investigation.

ii. The water level goes down in the water tank because there is a hole in the bottom of the tank (age 6 years)

A group of six year olds thought that there might be a hole in the bottom of their water tank which was the reason for the water level getting lower. Their teacher lifted up the tank for them to look underneath and they also felt the surface with their hands to see if it was wet. Having found no hole the children still wanted to check the joins at the edges of the tank. They decided to fill the tank up and to look later to see whether any water had come out. One child was concerned that the water might come out and then 'dry up' before they checked, so as a control they left some drops of water beside the tank. If these drops were still wet when the children came to check the tank then it was felt that any water which had leaked from the tank would still be there. The children came and checked later, and as their control drops were still there they were happy to accept that, since the table under the tank was dry, then no water was leaking out.

The children's revised suggestion was that a mouse had come and drunk the water. Although it might be argued that this hypothesis is not at a much higher level than that rejected, it is again testable. It is also an excellent example of how 'knowledge' must be discarded as well as constructed in the process of development.

iii. Some coffee goes into the air with the water from a coffee solution (age 10-11 years)

The children had noticed that the appearance of the coffee before making it into a solution with water, and again after the water had evaporated, was very different. It looked as if some coffee had 'gone' with the water when it evaporated, so the children decided to accelerate the evaporation process to see if they could locate the missing coffee. By heating the coffee solution, and condensing the steam on a mirror, they planned to test the condensed water to determine whether any coffee was present. The initial reaction to the appearance of the water was that the absence of a brown colour suggested no coffee was present. Testing by taste was suggested in order to corroborate observations. However, the interaction between taste and smell caused problems since the aroma of coffee was very noticeable and the children could not differentiate this from the sensations associated with the taste of the liquid. A further test was devised, involving weighing the coffee before adding water to it and after all of the water had evaporated. This test seemed more conclusive to the children.

Fig. 4.5

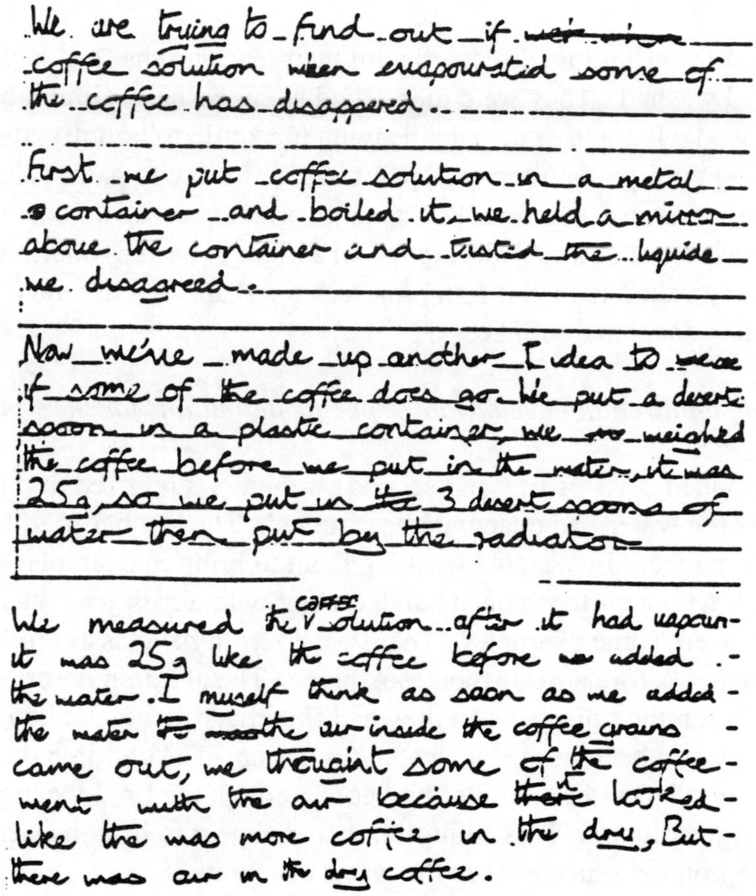

We are trying to find out if coffee solution when evapourstid some of the coffee has disappered.

First we put coffee solution in a metal container and boiled it we held a mirror above the container and tasted the liquide we disagreed.

Now we're made up another I idea to see if some of the coffee does go. We put a desert spoon in a plastic container, we weighed the coffee before we put in the water, it was 25g, so we put in the 3 desert spoons of water then put by the radiator.

We measured the coffee solution after it had evapour- it was 25g like the coffee before we added the water. I myself think as soon as we added the water the air inside the coffee grains came out, we thought some of the coffee went with the air because there looked like the was more coffee in the dry, But there was air in the dry coffee.

(Age 10 years)

"We are trying to find out if coffee solution when evaporated some of the coffee has disappeared.

"First we put coffee solution in a metal container and boiled it. We held a mirror above the container and tasted the liquids. We disagreed.

"Now we've made up another idea to see if some of the coffee does go. We put a dessert spoon in a plastic container. We weighed the coffee before we put in the water, it was 25g, so we put in 3 dessertspoons of water then put by the radiator."

"We measured the solution after it had evaporated. It was 25g like the coffee before we added the water. I myself think as soon as we added the water the air inside the coffee grains came out, we thought some of the coffee went with the air because it looked like there was more coffee in the dry, but there was air in the dry coffee."

b. Encouraging children to generalise from a specific context to others through discussion

A class of second year juniors discussed with their teacher what had happened to the clothes they had washed. They were then asked to suggest any similar happenings and they produced a list which included leaving the kettle to boil dry, drops of water left in a paddling pool going overnight, paint becoming dry in paint palettes. Interestingly this exercise also revealed some incorrect generalisations: milk 'drying up' into breakfast cereals, water 'drying up' in a plant pot after watering a plant. This exercise was very valuable in clarifying for both pupil and teacher the events they considered to be conceptually linked.

c. Encouraging children to develop more specific definitions for the words they use

A vertically-grouped class of first and second year juniors referred to the sugar and coffee left after the water had evaporated as 'stains'. The teacher tried to establish what the children meant by a stain by asking them to bring in examples of stained objects. A sock with a coffee stain, a handkerchief with a fruit juice stain and a tissue with a tea stain were some examples. These were left in the classroom for a week for children to observe before a discussion took place. The children described the differences between the objects when they had first made stains on them, and the objects after they had been in the classroom for a time. The children decided that a stain was what was left after the water had gone, though the final location for the water was not agreed upon. This definition would appear to be consistent with the solute remaining on the saucers when the water evaporated.

d. Finding ways to make the imperceptible perceptible

A group of lower junior children was shown a water atomiser, the spraying of which made water appear as small droplets which then ceased to become visible in the air. This activity was exciting for the children to watch and elicited comments concerning 'steam', 'clouds' and 'rain'. There were indications that this activity could be valuable as a means of extending understanding if presented at the appropriate stage in children's thinking, when they were able to appreciate the transformation of water from a perceptible to an imperceptible form.

> *e.g. "It's like a boiling kettle...steam...disappears into the air...hot water changes into drops of air".*

Summary

This Intervention work was found very challenging by the teachers. The response of the children indicated that there was an appropriate level at which this concept could profitably be addressed and the children pursued their activities with interest despite some initial teacher concern that the work would be too abstract for the pupils. The necessity of ensuring scientific rigour in the children's investigations was unfamiliar to some teachers and this led at times to children performing tests which confirmed rather than challenged their original notions. However, the increased awareness of the teachers of the importance of science process skills in concept development was a very positive outcome of the Intervention.

5. THE EFFECTS OF INTERVENTION

CHANGES IN CHILDREN'S IDEAS

INDIVIDUAL INTERVIEWS: Pre- and Post-Intervention

5.0 Introduction

As indicated in earlier sections, the work which was orchestrated by teachers in their own classrooms generated an enormous amount of qualitatively rich information about children's thinking. The classroom viability of the procedures developed was an important touchstone, as it was anticipated that the work would have relevance to, and might be replicated by, a wider circle of teachers at some point. The very strength of classroom viability also carries certain hazards from the point of view of collecting research data. For teachers, there are other curricular and 'house-keeping' demands; there is never enough time to examine all children's ideas either in breadth or in depth, and there is always the possibility of some ideas being overlooked. It was the intention that a more tightly controlled interview sample would enable a more confident commentary on the distribution of ideas to be offered.

A random sample of children was selected from class-lists provided by teachers. The sample was stratified so that equal numbers of boys and girls would be represented and equal numbers also from each of three achievement bands. Teachers were asked to allocate each child to a notional grouping based on judgements of overall scholastic achievement. This sample was then interviewed before and after the Intervention period as far as possible. The reality was that the interviewing was

a) time-consuming and labour-intensive;

b) possible only within a limited 'window' of opportunity; and

c) conditional upon the children having engaged in the relevant activities.

A target sample size of about 20 children in each of the age bands, 'Infant', 'Lower Junior' and 'Upper Junior' was the aim, with some redundancy built in to cater for unavoidable absences of children, teachers or interviewers. In the event, more of the post-Intervention interviews failed to materialise than had been anticipated. Wherever data may cast some light, however incomplete, they are nevertheless reported; where it seems helpful to do so, the age groups have been combined where numbers are low.

A point must be made about the interpretation of the data presented in the following pages in the light of the descriptions of children's thinking presented in Section 3. Both sections are drawing on the same schools and classes, the same teachers and activities and the same pool of children. The development of the categorisation systems for examining the distribution of children's ideas was an iterative process. All the experiences, classroom-based, from teacher meetings and from individual interviews, have contributed to the particular way of making sense of the data which is presented in the following pages.

In the following sections, summaries of children's ideas in five areas are presented and discussed:

5.1 Evaporation from a Tank of Water

5.2 Clothes Drying

5.3 Handprint on a Paper Towel

5.4 Coffee Solution

5.5 Sugar Solution

The activities drawn upon 'worked' in the sense that they have the potential to inform us as to how children are thinking.

5.1 Evaporation from a Tank of Water

For the purposes of this research, the water in the tank can be regarded as 'pure'; the activity of observing and accounting for the fall in water level is also pure in the sense of offering the most direct avenue of exploring children's ideas, uncontaminated by the more obvious context effects of some of the other activities.

The issues which will be examined in this section are the following:

Vocabulary used to describe the Process of Evaporation

Assumed location of the Missing Water

Agents of Change affecting the Water Level

Reversibility of the Process of Evaporation

Assumptions about the Physical State of the Missing Water

Nature of the Transformation of Water in the Process of Evaporation.

5.1.1 Vocabulary Used to Describe the Departure of Water

Table 5.1 summarises the main action verbs used by children to describe what happened to the water; the categories are not mutually exclusive as many children used more than one word. The non-technical words, 'disappeared' and 'vanished' together accounted for the most frequent descriptions of the non-presence of water. These words can be ambiguous. They are rarely used in a literal sense of 'dematerialisation' (though there are some instances in which this sense was intended). 'Disappeared' and 'vanished' often seemed to be used to describe an end state ('not visible') rather than a mechanism. Similarly, compounds such as 'gone up/ to/down' tended to be used to describe locations rather than mechanisms.

Just as the meaning intended by non-technical words was ambiguous, the use of 'vaporised' or 'evaporated' also required probing. Many children used both 'evaporated' and 'disappeared' in their descriptions.

It is not apparent that the Intervention had any impact on the vocabulary used in this instance.

Table 5.1 Action Verbs Used to Describe what Happened to the Water that was Missing from the Tank (Percentages)

	Pre-Intervention			Post-Intervention	
	Infts. n = 17	L.J. n = 18	U.J. n = 23	Infts & L.J. n = 13	U.J. n = 14
Evaporated/ Vaporised	6 (1)*	11 (2)	70 (16)	8 (1)	43 (6)
Disappeared/ Vanished	88 (15)	100 (18)	91 (21)	100 (13)	93 (13)
Gone up (to)	29 (5)	56 (10)	39 (9)	62 (8)	21 (3)
Gone down	53 (9)	39 (7)	48 (11)	15 (2)	14 (2)
Gone	-	6 (1)	9 (2)	-	7 (1)
Dried up	18 (3)	-	13 (3)	31 (4)	-
Dissolved	-	-	9 (2)	-	7 (1)
Melted	12 (2)	6 (1)	4 (1)	15 (2)	-
**Spilled/Taken	24 (4)	-	-	-	-
Other	-	-	13 (3)	8 (1)	-
No Response	12 (2)	-	9 (2)	8 (1)	7 (1)

* Raw figures are shown in brackets.

** Included in this category are responses using other words implying that a human or animal agent was responsible for the fall in water level.

5.1.2 Assumed Location of the Missing Water

It was not always possible to be clear as to what ideas children held about the final location of water. For some children, it was sufficient to say that the water had 'dried up'. Even after probing, there were occasions when further elaboration was not forthcoming; the expression was felt by some to be a complete and sufficient explanation, and whether or not the water was deemed to continue to exist remained an unknown.

A small number of children made it clear that, for them, the missing water indeed had ceased to exist. A slightly different response was puzzling to interviewers. This was the suggestion that the missing water had gone to the bottom of the tank. There were children who had difficulty with this question who suggested that the water level had gone down towards the bottom of the tank, but these were identified as a separate group. Since the interviews were conducted, the idea of the water 'contracting' (almost the idea of congealing) has been encountered elsewhere. Once again, this is a minority idea.

Another type of response, not very widespread but identified in particular with younger children, was the idea of leakage of a relatively local kind, to the shelf on which the tank stood, or to the ground. To some children who had encountered the idea of drainage and water pipes, the link seemed almost compulsive. (With hindsight, the location of the water tank on a shelf above the sink in one school, though convenient, was a distraction.)

A small number of children described the water as going to the sun; with increasing age, this type of response appeared to decrease. It suggests a confusion or over-identification of the agent with the location.

The only significant shift to be seen in Table 5.2 is the greater number of children amongst the upper juniors, post-Intervention, identifying the air as the final location of water from the tank.

Table 5.2 Ideas about the Location of the Missing Water (Percentages)

	Pre-Intervention			Post-Intervention	
	Infts. n = 17	L.J. n = 18	U.J. n = 23	Infts & L.J. n = 13	U.J. n = 12
Gone...					
to the air	-	28 (5)	26 (6)	15 (2)	50 (6)
to the sky	-	6 (1)	4 (1)	-	-
to the clouds	-	17 (3)	48 (11)	46 (6)	17 (2)
to the ground, shelf, drain etc.	24 (4)	6 (1)	-	23 (3)	-
to the sun	24 (4)	17 (3)	9 (2)	15 (2)	8 (1)
to the bottom of the tank	-	17 (3)	-	8 (1)	-
Unspecified, e.g. 'dries up'	30 (5)	-	-	8 (1)	17 (2)
Specific suggestions that water no longer exists	-	-	13 (3)	-	8 (1)
Other	12 (2)	-	-	-	-
No response	12 (2)	6 (1)	0	-	-
Don't know	-	6 (1)	4 (1)	-	-

5.1.3 Agents of Change Affecting the Water Level

Children were invited to comment on what they thought had made the water go from the tank and seemed to feel themselves to be on more confident ground in their responses than they had been in attempting to explain where the water had gone. As Table 5.3 indicates, most children mentioned either heat or air/air movement in some form. A source of heat was more frequently mentioned than the condition of access to, or movement of, air as a contributory factor.

Pre-intervention, no children mentioned both heat and air; post-Intervention, three of the thirteen upper juniors (23%) did so.

Table 5.3. Reference to Heat and Air as Agents of the Change in Water Level (Percentages)

	Pre-Intervention			Post-Intervention	
	Infts. n = 14	L.J. n = 17	U.J. n = 23	Infts & L.J. n = 10	U.J. n = 13
Heat source	71 (10)	65 (11)	74 (17)	80 (8)	77 (10)
Air/Wind/ Draught, etc.	-	47 (8)	30 (7)	30 (3)	46 (6)
Either heat or air mentioned	71 (10)	82 (10)	100 (23)	90 (9)	77 (10)
Heat and air mentioned	-	12 (2)	-	-	23 (3)
Neither heat nor air mentioned	14 (2)	-	-	-	-

The more specific agents which children specified as being responsible for the change in water level are presented in Table 5.4.

Younger children were more inclined to specify the sun as an agent, while older children were more likely to abstract the property of heat. Radiators were specified as heat sources by a small number of children.

'Wind' or 'Air movement' was more frequently expressed by upper juniors, post-Intervention. A small number of children referred to properties of air other than movement, using terms such as the 'air pushes' or the 'air sucks' the water.

A small number of children referred to clouds as agents rather than locations, and pre-Intervention (but not post-) there were some references to human (caretaker) or animal agents removing the water.

Table 5.4 Specific Agents of Change in the Water Level (Percentages)

	Pre-Intervention			Post-Intervention	
	Infts. n = 14	L.J. n = 17	U.J. n = 23	Infts & L.J. n = 10	U.J. n = 13
Heat	43 (6)	41 (7)	74 (17)	20 (2)	62 (8)
Sun	43 (6)	59 (10)	22 (5)	80 (8)	15 (2)
Radiator	7 (1)	-	13 (3)	-	8 (1)
Wind	-	6 (1)	4 (1)	10 (1)	-
Air movement	-	18 (3)	13 (3)	10 (1)	39 (5)
Air pushing/ Sucking	-	12 (2)	13 (3)	10 (1)	8 (1)
Clouds	-	6 (1)	-	10 (1)	-
Human/Animal	14 (2)	-	-	-	-

5.1.4 Reversibility of the Process of Evaporation

Further insights into children's understanding of the process of evaporation were gained as the result of asking them whether they thought it was possible to get the missing water back again. The sample sizes on which the data in Table 5.5 are based are small (particularly post-Intervention) and yet an important trend is noticeable. Whereas pre-Intervention, about half the sample denied the possibility of the process being reversible, none took this view after the Intervention. Post-Intervention, a larger percentage of children overall appreciated that the process was reversible, the water returning as rain or condensation.

Table 5.5 Ideas about Getting the Water Back (Percentages)

	Pre-Intervention		Post-Intervention	
	Infts. and L.J. n = 14	U.J. n = 20	L.J. n = 6	U.J. n = 10
As rain	36 (5)	45 (9)	100 (6)	50 (5)
As condensation	-	5 (1)	-	33 (3)
Not possible	43 (6)	50 (10)	-	-
Don't know how	7 (1)	-	-	-
No response	14 (2)	-	-	17 (1)
Other	-	-	-	-

5.1.5 Assumptions about the Physical State of the Missing Water

Children were not directly asked to make explicit their assumptions about the physical state of the missing water, but their comments during the course of the interview often revealed their assumptions. Data are summarised in Table 5.6.

A small number of children seemed clear in their assertion that the missing water no longer existed. These children could be described as 'non-conservers'. (It is possible that this response would be encountered more frequently if the sampling were to be extended to pre-school children).

When children use terms such as 'vanished', 'dried up' or 'disappeared' without further elaboration, their understanding of the physical state of the missing water is best regarded as ambiguous.

The major response category was that including all responses which used only the term 'water' with no transformation suggested during the course of the interview. Pre-Intervention, this accounted for 62% of all responses (n = 55); post-Intervention, the proportion was 48% (n = 21), though it cannot be suggested with confidence that this was because suggestions about the physical state of the missing water had become more specific or differentiated.

A much smaller proportion of children than referred to 'water' suggested a transformation of some kind; some of these transformations were to a perceptible form, others to an imperceptible form. Though the numbers involved in each group are small, this distinction is justified on the grounds of an important difference in understanding implied by each.

The perceptible forms - drops, droplets, mist, fog, steam, cloud, etc. - reveal children's attempts to explain a difficult phenomenon by reference to aspects of their direct experience. Pre-Intervention, 16% of children (n = 55), none of them infants, offered this type of explanation. Post-Intervention, the figure was 24%.

The kind of response which referred to the physical state of the absent water as a gas or particles - gas, oxygen, hydrogen, particles, vapour, etc. - revealed an attempt at a mental model rather than dependence on a perceptual experience to support thinking. Incidence of this type of thinking was surprisingly high: 9% pre-Intervention (n = 55) with no infants offering this type of response.

Post-Intervention numbers are depleted, but the 9% level remains in evidence.

Table 5.6 Assumptions Revealed about the Physical State of the Missing Water (Percentages)

	Pre-Intervention			Post-Intervention	
	Infts. n = 17	L.J. n = 18	U.J. n = 20	Infts & L.J. n = 8	U.J. n = 13
No longer exists	-	-	4 (1)	-	21 (3)
Disappeared/ vanished	6 (1)	-	9 (2)	-	-
Dried up	24 (4)	-	-	-	23 (3)
'Water' only mentioned	59 (10)	56 (10)	61 (14)	50 (4)	46 (6)
Drops or droplets	-	17 (3)	-	25 (2)	-
Mist, fog or steam	-	6 (1)	17 (4)	13 (1)	15 (2)
Cloud	-	6 (1)	-	-	-
Gas, oxygen, hydrogen	-	-	4 (1)	-	8 (1)
Air or part of air	-	11 (2)	-	-	-
Very small particles	-	6 (1)	4 (1)	13 (1)	-
Vapour	-	-	-	-	-
No response	12 (2)	-	-	-	-

5.1.6 Nature of the Transformation of Water in the Process of Evaporation

As well as revealing something of their ideas about the physical state of the missing water, children also provided sufficient comment for inferences to be drawn about how they viewed the nature of the transformation which had taken place. Six categories were developed to describe the variety of transformations which children described. The categories are mutually exclusive in some respect, but can also be described in terms of three superordinate categories.

1. No Necessary Conservation

   This group, which is not sub-divided, includes all those responses which focus on the **remaining** liquid in the tank, or refer to a possible de-materialisation of the missing water, e.g. 'The water's gone down', 'It's dried up' or 'It's disappeared', etc.

2. Change of Location with No Physical Change in the Nature of the Water

   The second group includes three kinds of response which have in common the factor of relocation of water, but with no physical change in the nature of the water. All responses in this group imply a conservation of the water.

   The first sub-group contains those responses which designate a human or animal agent as being responsible for the water's relocation.

   The second sub-group assumes that a non-animal physical agent is responsible for the relocation, but the end point is the site of the agent itself. For example, the sun may be identified as the agent, and the water is described as moving to the sun itself.

   The third sub-group refers to a non-animal physical agent as being responsible for moving the water to a new location, other than the site of the agent itself.

3. Physical Change in Nature of Water Associated with Change of Location

   Responses of this type conserve water, i.e. nothing is lost, apparently, as the result of the process of evaporation. The two categories in this group share the characteristic of the nature of water being physically changed during the change of location. The distinguishing characteristic between the groups is that in one set of responses, children hold on to a notion of water in perceptible form, e.g. mist, fog, steam, spray, droplets, etc., while in the other set, an imperceptible change is accepted, e.g. water vapour, gas, 'thin air', etc.

The distribution of responses across these categories is summarised in Table 5.7.

Table 5.7 Nature of the Transformation of Water in the Process of Evaporation (Percentages)

	Pre-Intervention			Post-Intervention	
	Infts. n = 17	L.J. n = 18	U.J. n = 23	Infts & L.J. n = 13	U.J. n = 14
No necessary conservation: focus on remainder	35 (6)	-	13 (3)	8 (1)	29 (4)
Human or animal agent	18 (3)	-	-	8 (1)	-
Change of location to site of agent	29 (5)	22 (4)	4 (1)	-	14 (2)
Change of location by physical agent	6 (1)	33 (6)	57 (13)	38 (5)	36 (5)
Physical change in water, but perceptible form	-	28 (5)	17 (4)	31 (4)	14 (2)
Water changes to imperceptible form	-	17 (3)	9 (2)	15 (2)	17 (1)
No response	12 (2)	-	-	-	-

Pre-Intervention, none of the responses from infants invoked any transformation of water, and a minority of infants attributed the change in water level to human or animal agents.

Overall, the proportion of children suggesting that the water moves to the site of the cause of the fall in level (usually the sun, less frequently clouds) was 17% pre-Intervention (n = 58) and 7% after (n = 27).

CHANGES IN INDIVIDUAL CHILDREN

Figure 5.1 summarises the specific nature of the shifts in ideas made by the 27 individuals for whom complete pre- and post-Intervention interview data were available.

Thirteen individuals, 48% of the sample, were located in the same categories pre- and post-Intervention.

Figure 5.1: Shifts in Individual's Ideas Between the Major Categories (n = 27)

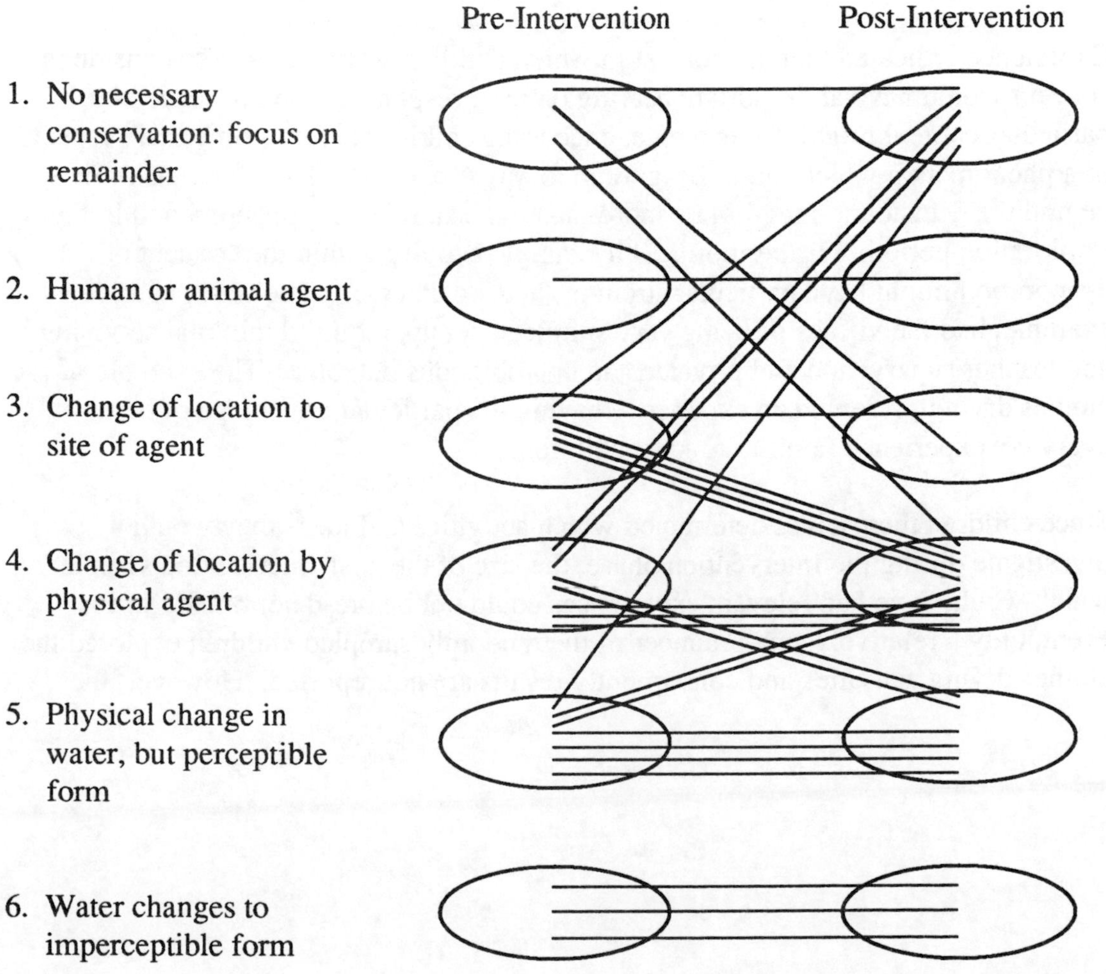

Pre-Intervention | Post-Intervention

1. No necessary conservation: focus on remainder
2. Human or animal agent
3. Change of location to site of agent
4. Change of location by physical agent
5. Physical change in water, but perceptible form
6. Water changes to imperceptible form

Most shifts were associated with the first category, which was defined to include those responses in which there was a focus on the remainder, or no necessary conservation. By definition, this category included the more ambiguous responses, so it is not too surprising that post-Intervention interviews resulted in a movement to another category. In four cases, children's post-Intervention responses dropped from qualitatively more elaborate responses; in two cases, responses moved from the first group to the third and fourth.

The other incidences of improved quality of response were four cases (15%, n = 27) shifting from a 'site of agent' to 'caused by agent' type of response. Two children (7%, n = 27) described water as being transformed in some perceptible way after the Intervention, in a way in which they had not before.

Again, it must be said that with complete information on such small numbers, it is risky to attempt to read too much from the data. What is clear, and encouraging, is that this systematic way of organising children's ideas offers a very useful support for making sense of their ideas about evaporation.

5.2 Clothes Drying

Experience indicated that the context in which children were required to consider a concept would have an important bearing on their response. The influence of a particular context might, for example, encourage children to focus on certain aspects of a phenomenon which might be ignored totally elsewhere. This thinking was behind the decision to incorporate more than one example of a phenomenon in the Exploration period. Furthermore, each example activity within the concept of evaporation might have particular strengths and weaknesses. The water tank, for example, had the virtues of being very simple to set up, required minimal servicing and teacher supervision and produced an unambiguous outcome. The example of clothes drying presented an experience having the particular advantage of being an everyday experience familiar to most children.

Since children themselves determined which activities and ideas they would investigate during the Intervention phase, the size of the post-Intervention sample which would have had relevant experiences could not be pre-determined. In the event, only a relatively small number of the randomly sampled children explored the clothes drying activities and consequently results are not reported. However, the pre-Intervention data provide some interesting points of comparison with responses to the water tank activity. The following areas are reported:

Vocabulary Used to Describe Clothes Drying

Final Location of Water from Drying Clothes

Drying Agents

Reversibility of Process

5.2.1 Vocabulary Used to Describe Clothes Drying

Table 5.8 summarises the action verbs used by children in describing the absence of water, once the clothes had dried on a clothes line. Note that many children used more than one of the relevant verbs.

Table 5.8 Action Verbs Used to Describe Absence of Water in Clothes Dried on a Line (Percentages)

	Pre-Intervention		
	Infts. n = 15	L.J. n = 16	U.J. n = 20
Evaporated	-	-	15 (3)
Disappeared	-	6 (1)	-
Gone up/down	7 (1)	25 (4)	20 (4)
Dried/dried up	47 (7)	31 (5)	65 (13)
Melted	-	6 (1)	-
Taken away	-	-	5 (1)
Blown away	-	13 (2)	10 (2)
Sunk in/soaked into material	-	6 (1)	15 (3)
Drips/drops/falls, etc. from material	67 (10)	63 (10)	45 (9)
Other	-	-	5 (1)
No response	13 (2)	-	-

In comparison with the responses offered to the question about what had happened to the water from the tank (Table 5.1), the profile of responses in response to the question about what had happened to the water from the clothes was very different. Perhaps because the tank presented a stronger visual event, words such as 'disappeared' and 'vanished' dominated children's descriptions. The most frequent description of the absence of water from clothes was that it had 'dried' or 'dried up'. We put clothes on the line to 'dry'. 'Drying' might also be thought of as essentially tactile, parallel to the visual information about changing water level in the tank. We test whether clothes have dried by gripping them, or holding them to the cheek, as well as by referring to their appearance.

The other major difference in comparison with the tank are responses (a) suggesting that the water has fallen from the material and (b) that the water has soaked into the material.

5.2.2 Location of Water from Drying Clothes

Children's ideas about the final location of the water from the drying clothes are summarised in Table 5.9.

Table 5.9 Ideas About Final Location of Water from Drying Clothes (Percentages)

	Pre-Intervention		
	Infts. n = 14	L.J. n = 16	U.J. n = 20
Unspecified e.g. 'dried up'	14 (2)	6 (1)	5 (1)
Net decrease in water	-	-	5 (1)
To sun/heat source	7 (1)	6 (1)	-
Into air/sky/ clouds/space	7 (1)	50 (8)	70 (14)
Into the ground, floor, etc.	64 (9)	19 (3)	11 (2)
Into the cloth/ soaked in, etc.	-	19 (3)	21 (4)
Other	7 (1)	-	-

When clothes are dried, it is common practice to wring as much water as possible from the material by hand before suspending them on a line with as large a surface as possible exposed. Water will usually drip from the material (especially synthetic 'drip-dry' materials). So three distinct processes are involved in drying clothes - physical compression, which is tangible; water falling under force of gravity, which is observable; evaporation, which is imperceptible.

It is interesting to note from Table 5.9 above that the younger children seem to have focused on the dripping effect, while the older children show awareness of the water leaving the material by some other process which is imperceptible. There is an additional response which is interesting in that it is offered only by the older children - that is, the water soaking into the material. There is possibly a confusion here with the function of absorbent paper towels and other materials used for soaking up water. (This point will be developed further in the discussion of the interview activity which utilised paper towels). A perspective on Table 5.9 is provided by a consideration of the perceptual experiences which children may have encountered in association with clothes drying. The idea of the water going to the air/sky/clouds/space shows a dramatic increase with age, just as the notion of the water falling to the ground decreases in dominance. The context of clothes drying made the idea of water going to the air more immediate or acceptable than in the context of the water tank. (Though after the Intervention, many more children used this explanation for evaporation from the water tank; see Table 5.2).

5.2.3 Drying Agents

In this section, children's ideas about the part played by evaporation and gravity in drying clothes, and then more specifically, the identification of heat and moving air as factors in evaporation, are examined.

In Table 5.10, the incidence of references to the idea of gravity and evaporation is summarised. These references were not literal; 'evaporation' responses include those in which children suggest that the water from the drying clothes went into the air, while 'gravity' responses referred to water dripping, falling to the ground, etc.

Table 5.10 Incidence of 'Gravity' and 'Evaporation' Response Types (Percentages)

	Pre-Intervention		
	Infts. n = 14	L.J. n = 16	U.J. n = 20
'gravity' **and** 'evaporation'	7 (1)	31 (5)	35 (7)
'evaporation' but **not** 'gravity'	-	25 (4)	40 (8)
'gravity' but not 'evaporation'	64 (9)	31 (5)	10 (2)
Neither 'gravity' nor 'evaporation'	14 (2)	13 (2)	15 (3)
No response	14 (2)	-	-

The high incidence of 'gravity'-type responses from younger children is particularly noticeable, though this as the sole explanation declined steeply through the age groups. The infants did not offer 'evaporation'-type responses as a sole explanation, though the junior children did. Of the upper junior age group, 75% mentioned either evaporation or gravity; 35% mentioned both.

Setting aside the issue of water dripping or falling from the wet clothes, Table 5.11 summarises the recognition of heat and/or air as a necessary condition in the process of clothes drying.

Table 5.11 Heat and Air as Agents of Clothes Drying (Percentages)

	Pre-Intervention		
	Infts. n = 10	L.J. n = 13	U.J. n = 19
Heat mentioned	70 (7)	69 (9)	47 (9)
Air/draught mentioned	40 (4)	62 (8)	63 (12)
Both heat and air mentioned	30 (3)	39 (5)	11 (2)
Either heat **or** air mentioned	50 (5)	54 (7)	90 (17)
Neither heat nor air mentioned	-	-	-

About 50% of children in the two lower age groups mentioned either heat or air as playing a role in the process of clothes drying. The proportion of juniors referring to heat is relatively low. How is the relative neglect of the condition of heat by juniors to be interpreted? It might be that younger children are unnecessarily pre-occupied with tangible radiant heat - from the sun or from radiators - while the older children are beginning to acknowledge that heat is not essential. That is, the clothes can dry in cold conditions, provided there is air movement (and low humidity).

5.3 Handprint on Paper Towel

The evaporation of water from an absorbent paper towel has some superficial similarities to the evaporation of water from drying clothes. An important difference is that the hand-print is clearly visible initially, enlarges as water is absorbed and then gradually the shape becomes less discernible as water evaporates. All this happens in less than two minutes provided the air contact is optimised, so the perceptual experience has some unique features.

5.3.1 Vocabulary Used to Describe the Paper Towel Phenomenon

Table 5.12 summarises the action verbs which children used when asked to describe what had happened to the handprint. The largest category of response was that describing the water as having 'gone' or 'dried up'. If the 'disappeared' group is added to these, it will be appreciated that pre-Intervention, the great majority of responses were couched in a rather generalised vocabulary. The exception are those responses using a vocabulary describing water as 'dripping' or 'falling from' the paper towel, and those describing water as 'sinking in' or 'soaking into' the towel. The 'dripping' responses only occurred in the two younger age groups; the 'soaking in' responses only occurred in the two older groups, pre-Intervention.

Table 5.12 Action Verbs Used to Describe the Absence of Water in the Paper Towel (Percentages)

	Pre-Intervention			Post-Intervention	
	Infts. n = 17	L.J. n = 18	U.J. n = 23	Infts n = 13	U.J. n = 14
Evaporated	-	6 (1)	-	13 (1)	9 (1)
Disappeared	6 (1)	13 (2)	10 (1)	-	9 (1)
Gone up/down	33 (6)	38 (6)	50 (5)	38 (3)	27 (3)
Dried up	61 (11)	44 (7)	60 (6)	38 (3)	46 (5)
Dissolved	-	-	-	-	-
Melted	6 (1)	-	-	13 (1)	-
Taken away	-	-	-	-	-
Blown away	6 (1)	-	-	13 (1)	9 (1)
Sunk in/soaked **into** material	-	25 (4)	40 (4)	13 (1)	55 (6)
Drips/drops fall, etc. **from** material	28 (5)	6 (1)	-	25 (2)	-
Other	-	-	-	-	-
Don't know	-	-	-	-	9 (1)

5.3.2 *Final Location of Water from Paper Towel*

Although the vocabulary used tended to be of a generalised nature, it was often possible during the individual interviews to probe children's thinking in order to ascertain more precisely what they had in mind. Table 5.13 summarises children's ideas about the final location of the water from the evaporated handprint on the paper towel.

Table 5.13 shows that the major response categories for the upper juniors are roughly equivalent to 'evaporation' (water goes to the air/sky/space/clouds) and 'soaking in'. Both of these types of response increase through the age groups. The other significant response group for the infants is the suggestion that the water goes to the floor, or to the ground - a response which seemed to decrease in frequency, post-Intervention.

Table 5.13 Ideas About Final Location of Water from Hand-print on Paper Towel (Percentages)

	Pre-Intervention			Post-Intervention	
	Infts. n = 18	L.J. n = 18	U.J. n = 9	Infts n = 8	U.J. n = 11
Unspecified e.g. 'Dried up'	6 (1)	-	11 (1)	38 (3)	9 (1)
To sun/heat source	11 (2)	6 (1)	-	25 (2)	9 (1)
Into air/sky/ clouds/space	11 (2)	33 (6)	56 (5)	13 (1)	46 (5)
Into the ground floor, etc.	28 (5)	6 (1)	-	13 (1)	-
Into the cloth/ soaked in, etc.	22 (4)	39 (7)	44 (4)	13 (1)	46 (5)
Other	6 (1)	-	-	-	-

5.4 Evaporation of Coffee and Sugar Solutions

Children's ideas about what happened to the coffee and sugar solutions are reported separately because the results imply differences in perception of the two events. In the generalised view, the evaporation of water from a solution of coffee and from a sugar solution are two examples of a single phenomenon. For children in the primary age range, there were two important differences, one of language use, the other perceptual.

In everyday language, there is a clear distinction between 'sugar' and 'sugar solution'. This is not the case with coffee; we use 'coffee' to make a cup of 'coffee'. Since the interviews with children were designed to extricate ideas about what happened to the solid and liquid components respectively, this confusion of language posed real problems. In fact, the language usage seemed to shape children's ideas about the process of the change also, in the sense that there was some idea of coffee having a 'natural state', that natural state apparently being liquid.

The particular perceptual impact which was realised to be important related to sugar solution and the way in which

a. sugar changes from opaque to translucent to transparent as it is added to water;

b. crystalline sugar tended to form a crust over some remaining solution in the saucer;

c. the sugar solute takes on a glassy appearance once the water has evaporated, or (more hazardous conceptually), the appearance of ice. This analogy with ice is supported when it is noted that some water remains below the surface crust. The perceptual similarities with ice can perhaps lead children to assume that the glassy crystalline sugar contains water. In contrast, the remaining coffee solids appeared virtually two-dimensional, like a trace of pigment. Furthermore, the glazed appearance of the coffee solute was very different to the light granular texture of the starting condition.

5.4.1 Coffee Solution: 'What's Gone?'

Most children were of the view that all of the water but none of the coffee mixture had gone from the saucer of coffee solution. About a quarter of the whole sample (24%, n = 37, pre-Intervention) suggested that coffee solids had gone as well.

Only fourteen children, all in the junior age range, were involved in Intervention activities associated with the coffee solution; post-Intervention, 21% of these mentioned that some coffee had gone along with all the water. Table 5.14 summarises responses.

Table 5.14 Coffee Solution: 'What's Gone? (Percentages)

	Pre-Intervention			Post-Intervention
	Infts. n = 10	L.J. n = 8	U.J. n = 19	L.J. & U.J. n =14
All water, All coffee	10 (1)	-	5 (1)	-
All water, Some coffee	20 (2)	-	5 (1)	21 (3)
All water, No coffee	70 (7)	88 (7)	53 (10)	64 (9)
Some water, Some coffee	-	-	16 (3)	-
No water, Some or all coffee	-	13 (1)	-	-
No water, No coffee	-	-	16 (3)	-
Other	-	-	5 (1)	7 (1)
Don't know	-	-	-	7 (1)

5.4.2 Final Location of Water from Coffee Solution

The most frequent response from the juniors as to where the water from the coffee solution had gone was of the kind, "to the air/sky/clouds", (50% of the 27 upper and lower juniors). Only one of the eleven infants interviewed offered this type of response in suggesting the clouds as the final location.

Evaporation of water from a solution produced some ideas which had not been seen in quite the same form in other contexts, though they were most akin to the notion of water 'soaking in', as seen with the drying clothes and paper towel. Eleven children (29% of the 38 interviewed pre-Intervention) suggested that the water had gone into the saucer or the coffee; in its most explicit form, the water was described as having gone 'hard'. This type of response will be discussed further in connection with the sugar solution.

Table 5.15 Final Location of Water from Coffee Solution (Percentages)

	Pre-Intervention			Post-Intervention
	Infts. n = 11	L.J. n = 8	U.J. n = 19	L.J. & U.J. n =14
To the air	-	25 (2)	16 (3)	50 (7)
Sky	-	13 (1)	-	7 (1)
Clouds	9 (1)	13 (1)	42 (8)	7 (1)
Below saucer/ ground, etc.	36 (4)	-	-	-
Sun/heat/source	27 (3)	-	5 (1)	14 (2)
Into saucer	18 (2)	13 (1)	-	-
Into coffee	18 (2)	-	21 (4)	14 (2)
Hard water on saucer	-	13 (1)	5 (1)	-
Unspecified e.g. water 'dries up'	9 (1)	13 (1)	5 (1)	28 (4)
Specific suggestion that water no longer exits	-	-	5 (1)	-
Don't know	18 (2)	-	-	-
No response	-	13 (1)	-	-

5.4.3 Coffee Solution: Agent or Mechanism of Evaporation

When asked to explain what had caused the effect in the saucer, that is, 'What made it happen?', two-thirds of infants and three-quarters of juniors attributed the outcome to heat. A minority (but no infants) referred to the movement of air. Three juniors (12% of 26 children) referred both to heat and air movement.

Table 5.16 Coffee Solution: Agent or Mechanism of Evaporation (Percentages)

	Pre-Intervention			Post-Intervention
	Infts. n = 8	L.J. n = 8	U.J. n = 18	L.J. & U.J. n =13
Reference to heat source	63 (5)	75 (6)	78 (14)	77 (10)
Reference to movement of air	-	38 (3)	17 (3)	23 (3)
One of above	63 (5)	88 (7)	78 (14)	62 (8)
Both of above	-	13 (1)	11 (2)	23 (3)
Other than either of above	13 (1)	-	11 (2)	-
Don't know	13 (1)	-	-	-
No response	-	-	-	15 (2)

5.5.1 Sugar Solution: 'What's Gone?'

As with the coffee solution, the majority response was that all of the water but none of the solid had gone. Once again, 24% (9 of the 37 pre-Intervention interviewees) suggested that some or all of the sugar had gone along with the water.

Table 5.17 Sugar Solution: 'What's Gone?'

	Pre-Intervention			Post-Intervention
	Infts. n = 10	L.J. n = 8	U.J. n = 19	L.J. & U.J. n =14
All water, all sugar	20 (2)	-	-	7 (1)
All water, some sugar	-	-	5 (1)	7 (1)
Some or all water, no sugar	60 (6)	88 (7)	58 (11)	79 (11)
Some water, some sugar	10 (1)	-	16 (3)	-
No water, some or all sugar	10 (1)	13 (1)	-	-
No water, no sugar	-	-	16 (3)	-
Other	-	-	5 (1)	7 (1)
Don't know	-	-	-	-
No response	-	-	-	-

5.5.2 Final Location of Water from Sugar Solution

The suggestions offered as to the final location of the water are very similar to those put forward in the context of the coffee solution. Ten of the 38 children interviewed (26%) before the Intervention period suggested that the water went into the saucer, into the sugar or went 'hard', that is, solidified.

Table 5.18 Final Location of Water from Sugar Solution

	Pre-Intervention			Post-Intervention
	Infts. n = 11	L.J. n = 8	U.J. n = 19	L.J. & U.J. n =14
To the air	-	25 (2)	16 (3)	50 (7)
Sky	-	13 (1)	-	-
Clouds	9 (1)	13 (1)	42 (8)	14 (2)
Below saucer/ ground, etc.	36 (4)	-	-	-
Sun/heat source	27 (3)	-	5 (1)	7 (1)
Into saucer	18 (2)	13 (1)	-	-
Into sugar	9 (1)	-	21 (4)	14 (2)
Hard water on saucer saucer	-	13 (1)	5 (1)	-
Unspecified, e.g. water 'dries up'	-	13 (1)	5 (1)	36 (5)
Specific suggestion that water no longer exits	-	-	11 (2)	-
Don't know	18 (2)	-	-	-
No response	9 (1)	13 (1)	-	-

5.5.3 Sugar Solution: Agent or Mechanism of Evaporation.

Three-quarters or more of children in each age group referred either to a heat source or to air/movement of air as being responsible for the evaporation having taken place. No infants referred to air, and far fewer juniors mentioned air than mentioned heat. A minority of lower and upper juniors referred to both air and a heat source.

Table 5.19 Sugar Solution: Agent or Mechanism of Evaporation (Percentages)

	Pre-Intervention			Post-Intervention
	Infts. n = 8	L.J. n = 8	U.J. n = 18	L.J. & U.J. n =13
Reference to heat source	88 (7)	75 (6)	78 (14)	77 (10)
Reference to movement of air	-	38 (3)	17 (3)	23 (3)
One of above	88 (7)	88 (7)	72 (13)	62 (8)
Both of above	-	13 (1)	11 (2)	23 (3)
Other than either of above	13 (1)	-	11 (2)	-
Don't know	-	-	-	-
No response	-	-	-	-

6. SUMMARY

6.0 Summary

This report summarises the cycle of activities from the development of exposure activities in classrooms, through the elicitation of children's ideas and the various classroom Intervention strategies developed and implemented, and finally, the follow-up interviews. How is this work to be judged? Shifts in pupil ideas occurred in some areas and these are of interest. However, such a narrow criterion would overlook some more general aspects of this study which are at least of equal importance.

The SPACE Project could not have been sustained without the close collaboration which existed between the officers of the LEAs and the University researchers. The Head Teachers of all the schools involved were aware from the outset that the Project carried the full backing and commitment of their LEA. Teachers could identify a network of colleagues within their education authority as the result of the Project meetings, and the opportunity to share a common ground of experiences beyond normal classroom confines was grasped with enthusiasm. Collaboration with the researchers and advisory teachers, including visits to classrooms, also served to expose teachers to issues which might not otherwise have been addressed. The fact that a University award, the Certificate in the Advanced Study of Education, was also geared to the SPACE Project work (once again, supported by the LEA) served as a further motivation to those teachers who elected to pursue the programme. All these factors added up to a significant contribution to the development of the attitudes and practices of participating teachers. As a classroom-based research Project, SPACE had a significant INSET impact.

While the positive factors which contributed to professional development can be readily identified, more formally speaking, the Project's remit was to investigate children's thinking, not that of teachers. All those participating in the Project - not least the majority of teachers themselves - were aware of changes taking place in their attitudes, knowledge, skills and techniques of classroom management. However, neither these gains nor the doubts and struggles on the way to achieving them, were formally recorded. Managing the classroom Intervention strategies involved teachers in learning new practices through attempting to implement them. This process was likened to making a journey and simultaneously learning to ride a bicycle. As far as possible, there was someone alongside for support through the wobbles (this was one of the advantages of more than one teacher per school participating). There were alarming periods when some individuals felt a sense of being out of control, but there were clear signals of exhilaration also, when mastery was achieved and confident improvisation became possible. What this implies for the Intervention is that the exact nature of the children's experiences cannot be specified. Ideas were discussed with teachers, but the quality and quantity of the specific programmes which were devised as a consequence of these discussions were not closely monitored. Where teachers experienced particular difficulties, some aspects of the programme might not have been implemented at all. While the researchers took some pride in the fact that great care had been taken to avoid springing unexpected interviews and questions on unsuspecting children, it has to be admitted that some teachers were confronted with some very unfamiliar problems. For example, non-directive questioning and the management of investigatory science were novel techniques for some teachers.

For all the reasons outlined above, it would be unrealistic to regard the Intervention programme devised to help children to move their ideas in the areas of evaporation and condensation as optimal. Even those strategies which were adopted and implemented had not been proven in practice.

On the other side of the balance are the general positive issues which tend to enhance the programme of any social group which is in the limelight - the novelty of new contacts and ideas, the interest and attention of the LEA, possibly the kudos of the participation in the Project and the CASE award. These issues cannot be specifically weighted, but surely had some bearing on outcomes.

Returning to the core issues of the Project - the qualities and directions of children's thinking - some general outcomes will be summarised at this point. There are two main types of information which will be summarised; firstly, a bird's eye view, or map of the major features of the terrain which has been explored; secondly, some specific ideas and directions which became apparent in children's thinking.

Evaporation as Transformation

Evaporation and condensation, in scientific terms, are simply the processes of change of state of water from liquid to gaseous state and vice versa. What the SPACE research tells us is that the transformation is explained by children by means of one of a range of alternative but coherent ideas. These ideas are characterised particularly by four factors:

1. The conservation or non-conservation of water
2. Ideas about the change in location of water
3. Ideas about conditions or agents governing change of location
4. Ideas about the nature of the transformation of water.

With reference to these characteristics, it is suggested that any response offered by children concerning the evaporation of water can be interpreted in terms of one of the six groups below:

No necessary conservation	• A focus on the remaining liquid, rather than that which has evaporated. No necessary conservation of material; i.e. there is no necessity acknowledged of accounting for the missing water; it may be considered to no longer exist.
Change of location of water, but with no transformation; no necessary change of form	• Change of location by human or animal agent. • Change of location to the site of the agent of change (e.g. the sun causes water to move to the sun). • Change of location by an agent or condition in the physical world (e.g. heat, air movement, etc.).
Transformation of water and change of location	• Transformation of water to some other perceptible form enables change of location to take place (i.e. mist, spray, droplets, fog etc.) • Transformation of water to some other form which is imperceptible, (i.e. gas, vapour, tiny fragments, etc.) implying a mental construct.

Water Tank

For many children, especially the younger ones, 'dried up' was regarded as a sufficient explanation of what had happened to the missing water.

'Disappeared' and 'evaporated' were used with increasing frequency as children got older.

For a small number of younger children, the missing water ceased to exist. Some younger children thought that the water leaked from the tank to the ground, or to the drain. A small number of younger children suggested that water had been removed, either by a person or an animal.

The great majority of children did not discuss water as a substance having varying states. The minority of children recognising water as undergoing a transformation can be divided into those who referred to other perceptible forms as being intermediate (mist, spray, droplets, etc.) and those who describe imperceptible mental constructs such as 'gas', 'vapour' and 'particles'.

Some children, particularly younger ones, confused the agent of transformation with the final location of the water.

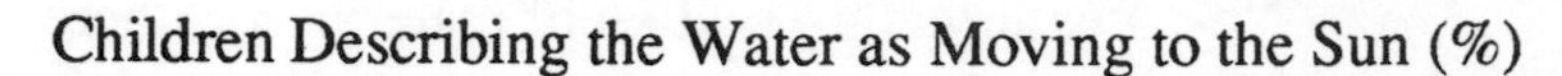

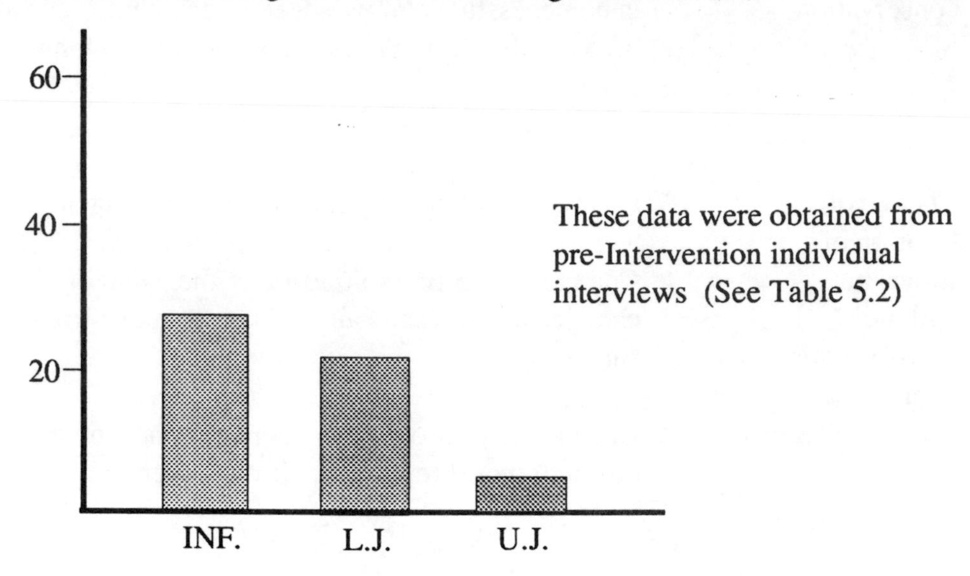

The only large shift in thinking as the apparent result of Intervention was the greater number of children amongst the upper juniors identifying the air as the final location of water from the tank.

Clothes Drying and Handprint on Paper Towel

Younger children tended to be content to use the expression 'dried up' as a sufficient explanation when asked to suggest where they thought the water had gone from the dried clothes.

Two descriptions of what had happened to the water which had not been relevant to the water tank, were used by children in relation to the drying clothes. These were 'dripping' and 'soaking in'.

'Dripping' was offered as an explanation by all age groups, but with much greater frequency by infants. Explanations involving water falling from the clothes were more likely to be offered as complete explanations by younger children.

Percentage of Children Explaining Clothes Drying
by Reference to Water Falling to the Ground.

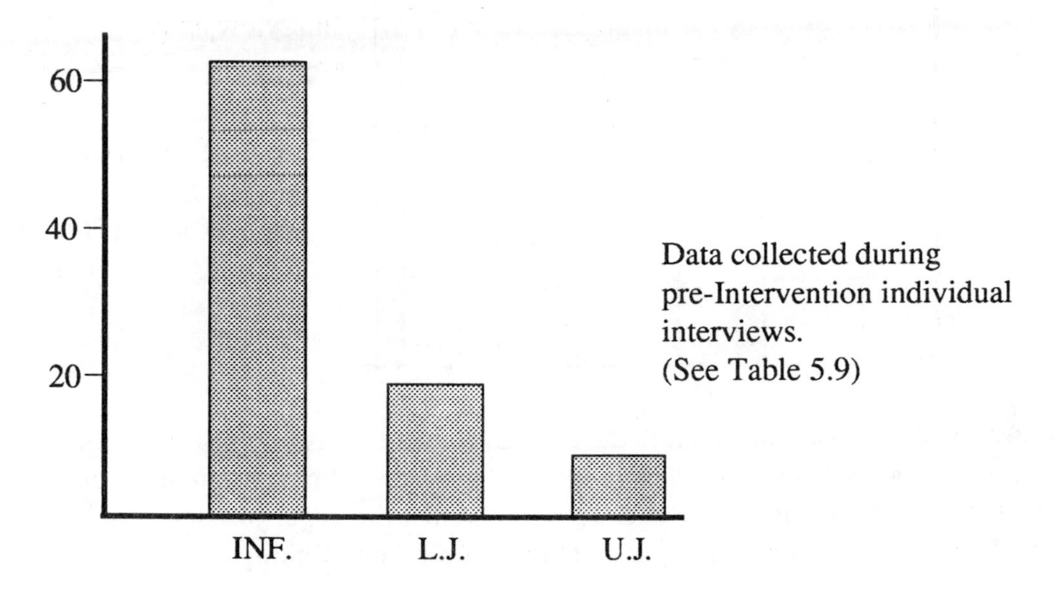

The 'soaking in' type of response seemed to increase with age. While it was not offered by any infants during individual interviews, it was suggested by about 20% of both lower and upper juniors. This trend was also seen in responses to the evaporation of the handprint from the paper towel.

Percentage of Children Explaining Evaporation of Handprint
from Paper Towel by Reference to 'Soaking In'

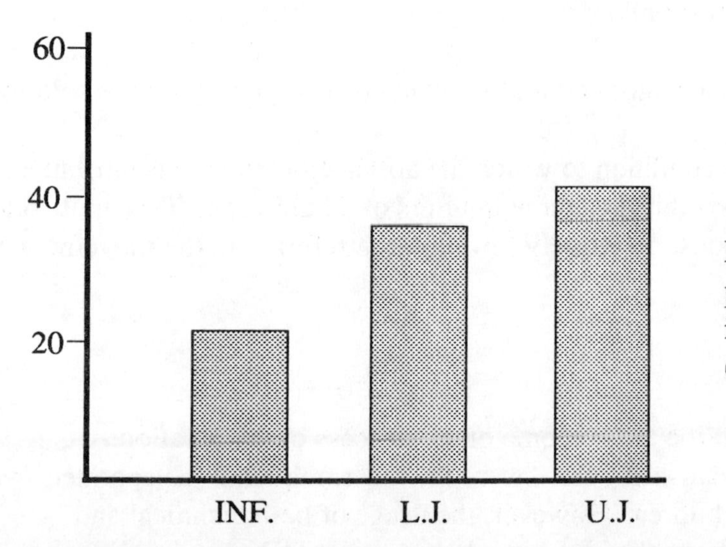

Data collected during
pre-Intervention individual
interviews.
(See Table 5.13)

The idea that the water from the drying clothes and from the handprint on the paper towel went to the sky was at much higher levels than had been the case with the tank of water. Furthermore, this type of response increased dramatically with age.

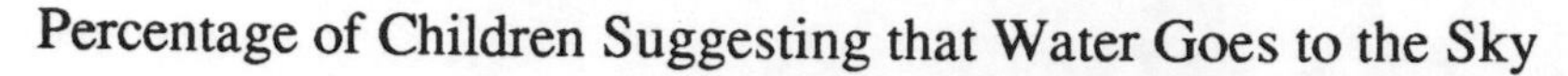

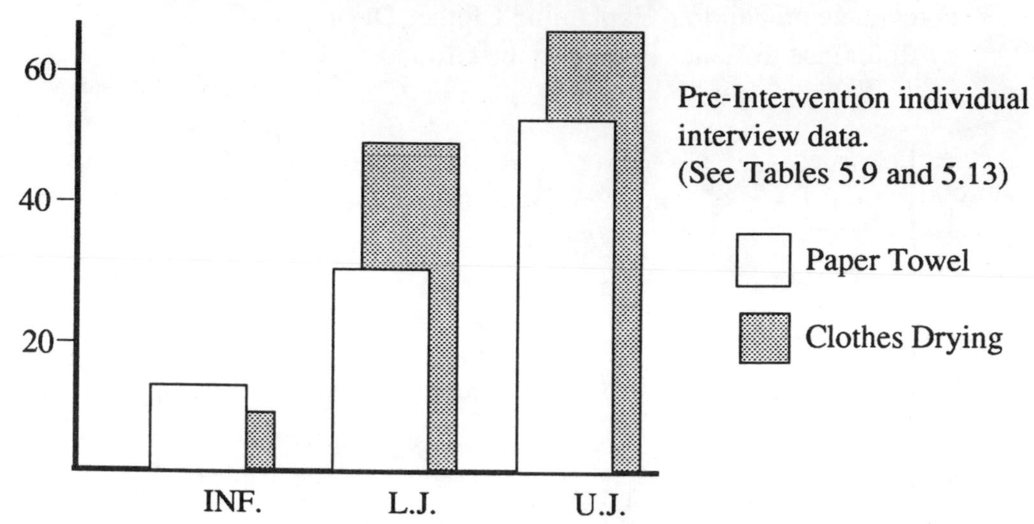

Children's ideas about the agent in the drying of clothes was interesting, in that the mention of heat as a condition decreased with age, while the mention of air or movement of air increased with age. It might be that the upper juniors were less dependent on the tangible quality of heat to support their thinking.

Evaporation from Solutions

More than half of the juniors described the water from solutions as going to the sky/ air/clouds.

More than a quarter of the children interviewed individually suggested that the water from the coffee or sugar solution went into the saucer, into the coffee or sugar, or simply went 'hard'. The notion of water as part of a solid compound other than ice, seems to be acceptable to some children.

About a quarter of all children suggested that some solids had gone from the solution.

Heat was the most favoured condition to which the absence of water was attributed, being suggested by about two-thirds to three-quarters of all children. This applied to both coffee and sugar solutions. Relatively few children referred to the movement of air.

Vocabulary

Non-technical words to describe the outcome of the process of evaporation - 'disappeared', 'dried up', 'vanished', etc. - were the most frequent. 'Evaporated' was used increasingly by older children. However, the usage of both technical and non-technical words had to be probed for intended meaning. 'Disappeared' was only rarely used in the sense of de-materialisation. 'Evaporation' was only rarely used with a correct understanding of what the word implies. Children frequently used 'disappeared' and 'evaporated' within the one explanation, often interchangeably.

Context Effects

Each of the main activities with which children were engaged presented slightly different facets of the evaporation process. In some ways, these activities (perhaps with additions) can be seen as complementary, both in eliciting children's view of evaporation, and at a later stage, in contributing to the development of their understanding. It is very unlikely that any single activity can serve either as a vehicle for eliciting ideas, or for developing them further. What is clear is that each activity brought particular aspects of the process to the fore. It should be possible to assemble these various attributes in the form of a more coherent and effective programme of classroom Intervention than has hitherto been possible.

BIBLIOGRAPHY

BAR, V. (1989): Children's Views about the Water Cycle. In: Science Education 73, 4, 483-500

BROOK, A., BRIGGS, H., DRIVER, R: Aspects of Secondary Students' Understanding of the Particulate Nature of Matter. CIS: Children's Learning in Science Project. Centre for Studies in Science and Mathematics Education. The University Leeds, January 1984
G6; P; AT

CHAMPAGNE, A.B., HALBWACHS, F., MEHEUT, M: Representations and Their Role in Learning in the Fields of Mechanics and Transformations of Matter. In: Research on Physics Education. Proceedings of the first international workshop. La Londe les Maures, 1983, 629-634
G6; C; P; M; AT

COSGROVE, M., OSBORNE, R.J: Physical Change. Working Paper 210. Learning in Science Project. University of Waikato, 1985, 28-57
G5; G6; P; M

DRIVER, R: Beyond Appearance: The Conservation of Matter under Physical and Chemical Transformations. In: Driver, R., Guesne, E., Tiberghien, A. (Ed.): Children's Ideas in Science, 1985, 145-169
G6; P; M; AT; C

KIRSCHER, E: Research in the Classroom About the Particle Nature of Matter (Grades 4-6). In: Proceedings of the international workshop on ''Problems Concerning Students' Representation of Physics and Chemistry Knowledge''. Ludwigsburg, 1981, 342-364
G6; P; AT

NOVICK, S., NUSSBAUM, J: Junior High School Pupils' Understanding of the Particulate Nature of Matter: An Interview Study. In: Science Education 62, 1978, 273-281
G6; P; AT

NOVICK, S., NUSSBAUM, J: Pupils' Understanding of the Particulate Nature of Matter: A Cross-Age Study. In: Science Education 65, 1981, 187-196
G6; P; AT

NUSSBAUM, J: The Particulate Nature of Matter in the Gaseous Phase. Driver, R., Guesne, E., Tiberghien, A (Ed.): Children's Ideas in Science
G6; G7; P; AT

OSBORNE, R.J., COSGROVE, M.M (1983): Children's Conceptions of the Change of State of Water. In: J. Res. Sci. Teaching 20, 9, 825-838

SERE, M.G: The Gaseous State. Driver, R., Guesne, E., Tiberghien, A. (Ed.): Children's Ideas in Science, 1985, 104-123
G6; G7; P; M

SHEPHERD, D.L., RENNER, J.W: Student Understandings and Misunderstandings of State of Matter and Density Changes. In: School Science and Mathematics 82 (8), 1982, 650-665

STAVY, R., STACHEL, D: Children's Ideas about ''Solid'' and ''Liquid'' European Journal of Science Education 7, 1985, 407-421
G6; P; M

Z'AROUR, G.I. (1976): Interpretation of Natural Phenomenon by Lebanese School Children. In: Science Education 60, 2, 277-287

APPENDIX I

Schools and teachers participating in the main experimental phases

KNOWSLEY

Senior Adviser for Science: Mr. B. Lord

School	Head Teacher	Teachers
St Andrews RC Primary	Mrs B Jones	Mrs P Deus Miss S McKenna
Park View Junior	Mr M J Adams	Mrs J Clarkson Mr B Dixon
Holy Angels RC Junior	Mr J E Yoxall	Mrs T Little Mrs D Standley
Brookside Primary	Mrs M Fisher	Mrs D Ellams Mrs M Smith
St John Fisher RC Primary	Mrs M Ledsham	Mrs A Donnelly Mrs D Martin Mr R Mason

APPENDIX II

PRIMARY
SCIENCE PROCESSES AND CONCEPT EXPLORATION
PROJECT

The SPACE project is funded by the Nuffield Foundation and is run jointly by Liverpool University, Centre for Research in Primary Science and Technology, and Kings College, London (KQC), Centre for Educational Studies.

The long term aims of this project are to influence the teaching of science concepts in primary schools. Certain key science concepts have been selected, and the initial task will be to establish what ideas children hold in these areas. Conversations with children often show that they try to make sense of the world around them by generalising from their own experiences. In this way, they develop coherent ideas which they use to explain things in their environment. However, these notions may be scientifically inaccurate and may actually hinder later formal learning of scientific concepts.

Talking to children will be centrally important in discovering what beliefs they hold in the chosen concept areas. Exposing the children to examples of a concept without giving any factual input should encourage them to think about the area. A variety of techniques, formal and informal, conducted in groups or individually, will then be used to explore their ideas.

Having explored children's informal ideas, the next step will be to consider ways of modifying them in the desired direction. The knowledge accumulated about children's starting points will form the basis for intervention. Children will be encouraged to test commonly recurring misconceptions during their own investigations. It remains to be seen whether children's ideas will be modified when they actively apply science processes in the chosen topic areas.

The flow chart on the opposite page details the activities which are envisaged up until December 1988.

FLOW CHART

Sept. to Dec.	1986	Detailed arrangements made with LEAs. Teachers identified; schools selected. Initial draft of concept list drawn up by project directors. Concept areas identified and initial tasks defined.
Jan 1st to Feb.	1987	Full-time researcher appointed. Meetings with teacher groups for discussion of draft concept list andassignment of activities.
March	1987	Teachers collect data using agreed activities.
April to May	1987	Data analysed by team.
June	1987	Evidence shared with teachers leading to suggestions of strategies for changing children's ideas. Teachers' opinions surveyed for the evaluation. Teachers prepare materials for activities for new class.
Sept. to Dec.	1987	Teachers present activities as before to children and collect information about existing ideas - this constitutes pre-intervention evaluation data. Follow up with teaching involving the intervention strategies agreed. Children's responses to application activities collected - this provides the post-intervention evaluation data. Teachers' opinions surveyed.
Jan. to Feb.	1988	Work on new concept areas begins and data collected - pre-intervention data on children (as in March 1987).
March	1988	Data analysed by team.
April to May	1988	Evidence to teachers, intervention proposed and implemented. Teachers' opinions surveyed.
June to July	1988	Children's responses to application activities collected. Post-intervention data on some pupils. Teachers' opinions also collected.
Sept. to Dec.	1988	Data interpretation and report writing.

A-3

Primary SPACE Topic Network

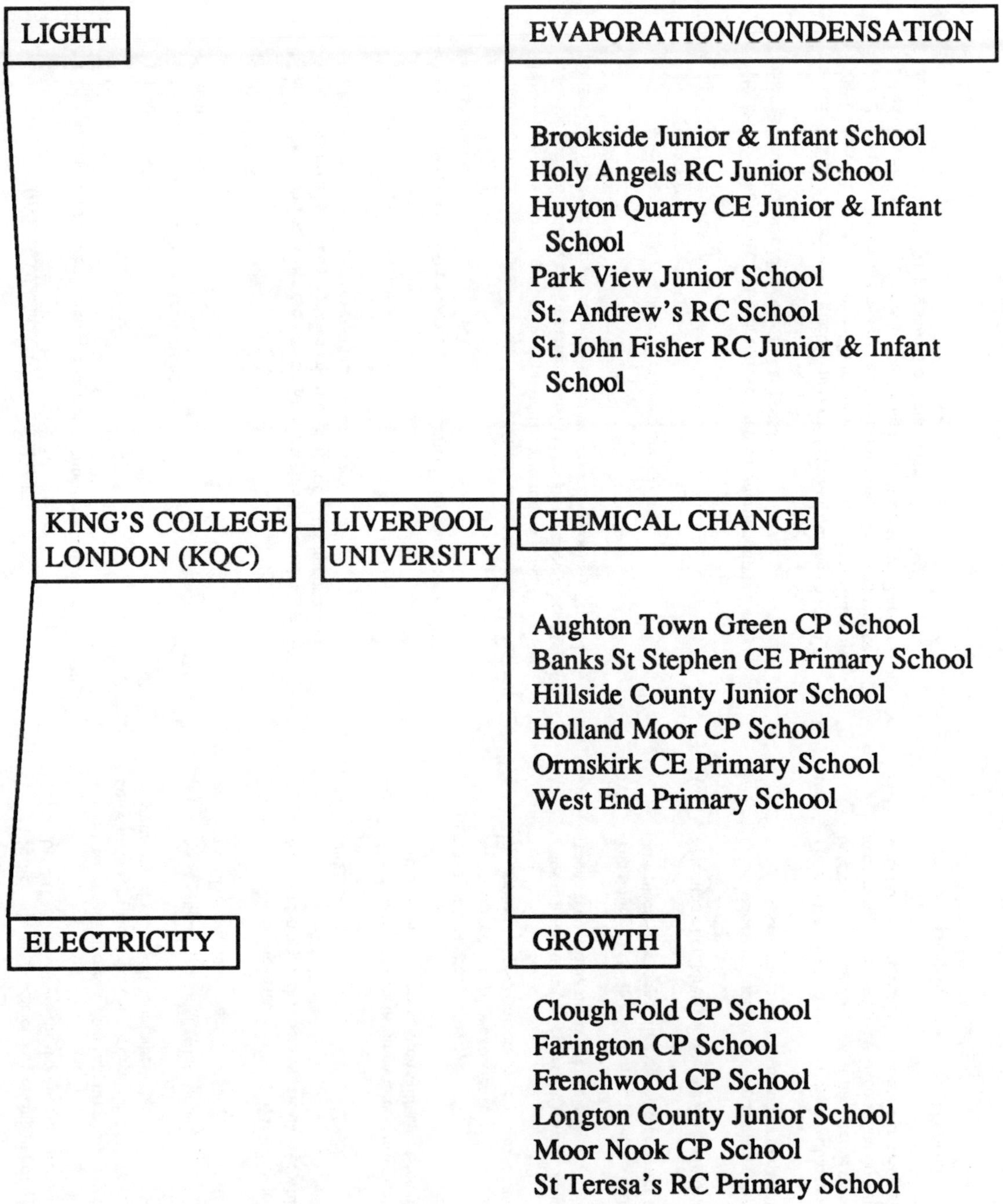

LIGHT

EVAPORATION/CONDENSATION

Brookside Junior & Infant School
Holy Angels RC Junior School
Huyton Quarry CE Junior & Infant School
Park View Junior School
St. Andrew's RC School
St. John Fisher RC Junior & Infant School

KING'S COLLEGE LONDON (KQC)

LIVERPOOL UNIVERSITY

CHEMICAL CHANGE

Aughton Town Green CP School
Banks St Stephen CE Primary School
Hillside County Junior School
Holland Moor CP School
Ormskirk CE Primary School
West End Primary School

ELECTRICITY

GROWTH

Clough Fold CP School
Farington CP School
Frenchwood CP School
Longton County Junior School
Moor Nook CP School
St Teresa's RC Primary School

DATA COLLECTION PROPOSALS

Useful information about children's concepts can be obtained in each of the following settings:

1. Class activities
2. Group activities within a class setting
3. Individual interviews

1. Class activities

(a) These background activities are set up by the teachers and are designed to expose children to the concept area under exploration. It is important that each class in a topic group covers the same activities.

(b) Each activity involves a change which occurs over a period of time.

(c) The activity is set up in a convenient place, either in the classroom or in a corridor area.

(d) The children are encouraged to make careful observations and record anything they notice in a 'diary', starting with an initial description. The 'diary' can take the form of writing and/or pictures, and observations should be at regular intervals. The time interval between observations will obviously depend on the activity, e.g.:

steel wool rusting - every 15 minutes initially
plants growing - every day
stick insects growing - once a week.

The children should not be led or directed in their interpretation of any changes, but encouraged to be as full as possible.

(e) A concluding, tape-recorded discussion is made to find out the children's ideas about the concepts involved in the activity. All questions which are asked by the teacher should be open and non-directing. Teachers will be asked to summarise the ideas emerging from these discussions. These summaries and labelled tapes should be kept in the buff folders.

2. Group activities

(a) These activities develop the ideas initiated in the background activities. They maybe conducted by the teacher in a way which is compatible with the classroom organisation.

Either i. the whole class working in groups at the same time
or ii. small groups working independently.

The teacher might also enlist the support of a group co-ordinator, research co-ordinator, or Head Teacher in conducting these group sessions. A written record or summary of ideas will be required.

(b) The children in a group might jointly record each activity in a relevant form, i.e. diagram, list, written account.

(c) The concluding discussion (1.(e)) is likely to tie in with the group activity since the group tasks focus on the change which has occurred.

3. Individual interviews

(a) These interviews focus on particular instances of the concept area under exploration. The orientation will have been provided by

i. a class or group activity
ii. assumed everyday experience
iii. a physical phenomenon or event based around (i) or (ii).

(b) They are conducted by either the teacher, head teacher, group co-ordinator or research co-ordinator, as available.

(c) The timing of the interview may coincide with the group activity or it may take place later, depending on the manpower available.

(d) Selection of children for interviewing will be on the basis of:

i. ideas produced in the group/class activities;
ii. approximately equal numbers of each sex.

(e) The number of children it will be possible to interview in detail will be determined by the duration of the interviews undertaken. Each group will need to make its own plans in consultation with the co-ordinators. The selection may be made by the teacher, head teacher or a co-ordinator.

(f) Each topic group of schools should provide at least 20 interviews per concept area for each year group, i.e.:

10 1st year interviews about evaporation;
10 1st year interviews about condensation
and the same for infants, 2nd, 3rd, 4th year juniors.

This may involve teachers/group co-ordinators in setting up materials in additional classes in the school, or placing them in an area accessible to other children, e.g. a corridor.

(f) The interviewer will record the child's ideas during the discussion with them, either by direct transcription or by later transcription if using an audio tape.

Data Collection

Any written material produced should be marked clearly with the date, the child's name (and sex if this is not obvious) and their year group. It should be kept in the buff folder.

INTERVIEWING TECHNIQUES

1. In order to explore children's own concepts it is important that their ideas should not have been influenced by teacher input. It is also important that the interviewer has no preconception about the content of the child's responses.

2. The child must be aware of the purpose of the interview, which is to establish the child's ideas about a particular topic. It must also be clear to the child that the interview is being recorded, either by transcription or on audio tape.

3. The interview questions in the topic pack are there as a guideline. One question may be enough stimulus for a whole interview. It may be more informative to follow the child's line of thinking than to try and go through the questions from beginning to end.

4. Any questions should be non-directive. They should be phrased so that the child can give any answer they feel is appropriate,

   e.g. "Tell me what you know about clouds"

5. The child should be allowed to talk freely, and the interviewer's attention given to picking up any critical words or phrases. These remarks can then form the basis of further questions which might focus more closely on the child's ideas.

   e.g. "We have to wash the beans twice a day because they need water or they'll shrivel up."

   "Can you tell me why you think the beans need water?"

6. The child should be encouraged to expand any answer they give, even if they appear to have contradicted previous statements. A look which registers the interviewer's surprise may inhibit the child's elaboration.

7. The questions should require no factual knowledge on the part of the child.

   e.g. "What is the name of this plant?"
   is not saying anything about the child's conceptual knowledge of plant growth.

   "What can you tell me about this plant?"
   might elicit the same information and also allow for other ideas to be put forward.

8. Questions with a yes/no answer should be avoided because they do not allow a free response.

   e.g. "Do you think the glass is cold?"
   could be better phrased as:

   "Have you any ideas why this happens to the glass?"

9. It is a good idea to repeat what the child has said to make sure:

   (a) that they have said what they meant;

   (b) that their intended meaning is what has been understood.

   Recapping may also prompt the child to elaborate their answer.

10. It is possible that the interviewer and the child may use the same word to mean different things. It is necessary to ensure that the child's everyday meaning is understood.

11. The length of the interview will obviously vary both with the subject matter and the child. It is important, though, that the discussion is not extended beyond its natural time limit in the hope of obtaining new ideas.

12. During the course of an interview, children may change their minds. When this happens, all the ideas expressed should be recorded, in sequence. The temptation to summarise the discussion in terms of the final idea expressed will give a false impression of commitment or certainty, and should be avoided.

13. All the above points stress the non-directive nature of the data collection. At the same time, interviewers must keep in mind the focus of interest of the interviews. This means:

    a) all the points of interest to the topic are covered;

    b) not too much time should be spent drifting into interesting but non-focal areas.

KIT LIST

EVAPORATION/CONDENSATION

Paper - diaries
Audio tapes
Cassette recorder
Paper towels
Hair dryer
Saucers
Jam jars
Large container to hold water
Mirrors
Waterproof pen/crayon
Cloths/sponges
Tins
Bread
Clothes/material
Clothes line/string
Pegs
Salt
Sugar
Coffee
Plant/mist spray
Biscuits
Plastic sheeting
Ice
Thermos flask
Paint trays
Paint
Paint brushes
Water
Rulers

Additional items may be required for group work

A Possible Schedule for Activities

TIMETABLE

WEEKS

	WEEK 1	WEEK 2	WEEK 3	WEEK 4	WEEK 5
	M.T.W.Th.F.	M.T.W.Th.F.	M.T.W.Th.F.	M.T.W.Th.F.	M.T.W.Th.F.
Class and group activities		large	container of	water	
	windows				
		paints			
		puddles			
			sugar		
			bread		
				clothes	
					cold air
Assessment of responses	—	—	—	—	—
Interviews	—	—	—	—	—

CLASS AND GROUP ACTIVITIES

These activities are ordered in terms of priority. Each school should agree to cover the same activities.

Evaporation

1. Drying out of bread

Kit: bread
rulers
paper
pens

Class

a) Put a slice of bread onto a piece of paper and draw round it.

b) Leave the bread until it is completely dried out (approximately 24 hours)

c) Draw round it again.

[Additional activity - remove the crust before leaving the bread to dry]

Group

a) Compare a fresh and a dried-out slice of bread.
Make a list of the things that are the same and different about them.

b) Do you think you could change the dried-out slice back to its original form? Try out your ideas.

Discussion

a) What is different about the two slices?

b) What do you think has happened to the hard slice?

c) Why do you think it has happened?

d) Do you think you could stop it happening?

e) Do you think you could change it back?

f) Does the same thing happen to anything else?

2 Container of Water

A Kit: large container (eg fish tank, bucket)
water
waterproof pen/crayon

a) Put water in a container almost to the brim and mark the water level.

b) Mark the level every day/week (depending on rate of evaporation).

B Kit: saucer/jar
water
sugar/coffee
waterproof pen/crayon

a) Dissolve sugar/coffee in some water.

b) Pour the solution onto a saucer, almost to the brim.

c) Mark on your saucer where the water comes to, then leave it somewhere safe. Watch it carefully during the day. Mark the water level at suitable intervals.

Discussion

a) What happens to the water?

b) Where do you think the water goes to?

c) Can you get the water back?

d) Does the same thing happen to every saucer?

e) What do you think is left on the saucers?

Class

Make some puddles in different parts of the playground and watch them during the day.

[Additional activities - make puddles on different surfaces]

Group

a) Where do you think the water goes to? Draw pictures to show your ideas.

b) Can you make the puddle last longer? Write down your ideas and then try them out.

c) Can you make the puddle go more quickly? Write down your ideas and then try them out.

Discussion

a) What happens to your puddles?

b) Does the same thing happen to all of them?

c) Where do you think the water goes to?

d) Can you make the puddle last longer?

e) Can you make the puddle go more quickly?

clothes line/string
pegs

Class

Wash (or simply wet) some clothes/material.
Peg them onto the clothes line/string, either inside the school or outside. Watch them periodically as they dry.

Group

a) Can you make the water go faster? Write down your ideas and then try them out.

b) Can you make the water last longer? Write down your ideas and then try them out.

Discussion

a) What happens to them while they are on the line?

b) Where does all the water go?

c) Can you make the water go faster?

d) Can you make the water last longer?

5 Drying-up of Paints

Kit: paint
paint trays

Class

Leave some paint in a paint tray to dry out. Look at it every day, and note down any changes you see.

Group

a) Do you think you can make the paint useable again?
Write down your ideas and try them out.

b) Do you think you can make the paint go more quickly? Write down your ideas and try them out.

Discussion

a) What do you think has happened to the paints?

b) Why do you think it has happened?

c) Do you think you could stop it happening?

d) Does the same thing happen to anything else?

Condensation

1. Breath on windows

Kit: windows

Breathe onto your classroom window.
Look carefully at what happens.

Discussion

a) What happens?

b) What do you think it is?

c) Where has it come from?

d) Can you make it go away again? Try out your ideas.

e) Where do you think it goes to?

2. Breath in cold air

Kit: a cold day!

Go outside and breathe out.
Look carefully at what happens.

Discussion

a) What can you see?

b) What do you think it is?

c) Where do you think it has come from?

d) What do you think has made it appear?

e) Where do you think it goes now?

f) Do you think you can make it go away?

g) Have you noticed anything like this happening anywhere else? (car exhaust, plane vapour trails, clouds, fog, mist.)

APPENDIX III

INDIVIDUAL INTERVIEWS

These activities may also be used with small groups.

Evaporation

1. Handprints

Kit: damp sponge/cloth
paper towels

Press your hand onto a wet cloth and make a handprint on a paper towel.

Wave the paper towel in the air.

Discussion

a) What happens to the handprint?

b) Where do you think the water has gone?

c) Do you think you could make the water come back?

d) How can you make the handprint last longer?

e) How can you make the handprint go more quickly?

2. Hand Drying

Kit: wet sponge/cloth

Wet your hands on a wet cloth and shake them in the air.

Discussion

a) What do your hands feel like when you shake them?

b) What happens to the water on your hands?

c) Where do you think the water goes to?

d) Can you make the water last longer? Try out your ideas.

e) Can you make the water go more quickly? Try out your ideas.

3. Paint Drying

Kit: paint
paint brush
paper

Paint a picture.

Discussion

a) Where are you going to put your picture to dry?

b) Why is that a good place?

c) What happens to your painting when it is drying?

d) Can you make your painting dry more quickly?

Condensation

1. A tin containing iced water

Kit: tins
water
ice

Put some ice and water into a tin.
Look carefully at the outside of the tin.

a) What do you think is on the outside of the tin?

b) Where do you think it has come from? Make a list of your ideas and try them out.

c) Do you think you can make it go away again? Try out your ideas.

d) Can you think of anywhere else you have seen this happen? Make a list.

2. Breath on a Mirror

Kit: mirror

Breathe onto a mirror.
Look carefully at what happens.

Discussion

a) What happens?

b) What do you think it is?

c) Where has it come from?

d) Can you make it go away again? Try out your ideas.

e) Where do you think it goes to?

Data Collection

The pilot phase in March 1987 provided us with a wealth of information covering the whole spectrum of children's ideas.This information has allowed us to target more accurately what it is we need to know from this Exploration phase, and from the intervention.

The essential difference between the pilot phase and Exploration is that the second is followed very closely by an intervention. For you, as teachers, to be able to influence your classes' ideas it is important for you to have a quick, easy way to know what their 'starting ideas' are. Diaries and interviews are very informative but the ideas need unravelling before they can be appreciated. We are suggesting, therefore, that there are two types of data collection: one rapidly accumulated and assessed and useful primarily for you (though we will also be interested in these findings) and the second more time-consuming both to collect and unravel, for which we will be responsible.

Classroom Data

1. A class 'logbook' into which children can write anything they wish to say about the activities. Each entry will need to have a name and date on it. Your skills in motivating the children will be important here, though it is not essential that every child makes an entry. The logbook is useful for seeing what particular children are taking an interest in, and for keeping children's attention on the activities.

2. A checklist of ideas. This list contains relevant questions and the corresponding ideas which we have found children to hold. This list is for your informal use, to give you an idea of how many different view points there are in your class, and of how prevalent they are. It is not intended that you start at the beginning and go through every question. Use it as a place to record what your class tell you rather than as a ready-made discussion with one question automatically following on from the previous one.

3. A small number of tasks, each designed to encapsulate as many of a child's ideas as possible on one sheet of paper. In order to make expression of these ideas as easy as possible for children each task can be done as a picture with explanatory labels. These labels can be added by the child him/herself or by you if the child tells you something pertinent. The pictorial nature of the tasks should free you to move round your class, discussing their pictures with them and helping them clarify any ideas which are ambiguous. To this end, you will also find within these papers is a list of words whose usage needs clarifying in discussion, eg. "disappeared". Obviously, if a child would rather write and is able to do so lucidly then that is equally useful. Each task is intended to take no more than about twenty minutes, though there is no upper or lower time limit imposed by us, and it is important that the task is not presented as a test; choose your moment well. You can, of course, use the checklist in combination with the tasks, to assess their content. We will be very interested in the ideas which emerge from each of these sources, and we would like to be able to remove them from school at the end of the term, along with any written comments which you have made.

Data from Selected Sample

Interviews will be conducted by members of the research team. As you all know, interviewing is very time-consuming and since you need to be aware of the views of all your children it is obviously not the best possible way for you to obtain your information. For our purposes, however, in-depth interviews are essential. In March we selected children to interview on the basis of their ideas, so that we could collect the whole range of ideas. Having done that, we are now able to be more systematic in our selection.

Selection for Interviews

We are asking you to group your class into three achievement bands; high, middle and low. Note, this is actual achievement in school subjects, not ability. We will then select children from each group to interview, and will interview up to twelve children per class.

Activities

These activities are intended to encourage the children to start thinking about the concept area under exploration. They are not meant to be a focus of learning, but should be kept as an interest point. This is important since they, or similar activities, will be central to study in the half term leading up to Christmas, and it will be detrimental to the intervention phase if the children's interest has been roused and lost too early in the proceedings.

The discussion questions which follow each activity are there for use if or when appropriate and they, like the tasks, should be kept very informal.

Some of the activities are accompanied by an 'exploratory task', an exercise which involves each child in putting their ideas down on paper in the form of an annotated diagram or writing. Some tasks are predictive, and so should be synchronised with the setting-up of the activity; others can be fitted in at a convenient time; others directly follow the completion of the activity.

Children who are interviewed may be asked about the class activities as well as, or instead of, an interview activity.

Kit List

*Large container (e.g. fish tank, bucket)
*Waterproof pen/crayon
Saucer/jar
Sugar/coffee
*Clothes/material
Clothes line/string
Pegs

*Not provided

EXPLORATORY TASKS

EVAPORATION/CONDENSATION

1. Activity - fish tank
   Timing - after two weeks

   Draw a picture of / write about the fish tank and show

   a) where you think the water has gone

   b) how you think the water got there

   Do you think you could make the water in the tank last longer/go faster?

2. Activity - clothes drying
   Timing - after washing, before drying

   Draw a picture showing /write about where you think would be the best place to put the clothes so that they will dry as quickly as possible. Also show what you think will happen to the water.

3. Activity - Breathing on a window
   Timing - directly following the activity

   Describe (in words or pictures) what you think is happening. Where do you think it comes from? Do you think you can make it go away?

ACTIVITIES - Evaporation/Condensation

1 Container of Water

Kit: large container (eg fish tank, bucket)
water
waterproof pen/crayon

a) Put water in a container almost to the brim and mark the water level.

b) Mark the level every day/week (depending on rate of evaporation).

Discussion

a) What happens to the water level?

b) Where do you think the water has gone?

c) What do you think has made the water go?

d) Do you think you can make the water go faster?

e) Do you think you can make the water last longer?

f) Do you think you can get the water back?

2 Solutions

Kit: saucer/jar
water
sugar/coffee
waterproof pen/crayon

a) Dissolve sugar/coffee in some water.

b) Pour the solution onto a saucer, almost to the brim.

c) Mark on your saucer where the water comes to, then leave it somewhere safe. Watch it carefully during the day. Mark the water level at suitable intervals.

Discussion (when all the water has evaporated)

a) Tell me what you think has happened.

b) Where do you think it has gone?

c) What do you think has made it go?

d) Do you think the same think happened to both saucers?

e) What do you think is left on the saucers?

f) Why do you think it has stayed behind?

3 Clothes drying

Kit: clothes/material
clothes line/string
pegs

Class

Wash (or simply wet) some clothes/material.
Peg them onto the clothes line/string, either inside the school or outside. Watch them periodically as they dry.

Discussion

a) What happens to them while they are on the line?

b) Where do you think all the water has gone?

c) What has made the water go?

d) Do you think you can make the water go faster?

e) Do you think you can make the water last longer?

f) Do you think you can get the water back?

4 <u>Breath in cold air</u>

Kit: a cold day! (or a window if the weather won't oblige)

Go outside and breathe out.
Look carefully at what happens.

Discussion

a) What can you see?

b) What do you think it is?

c) Where do you think it has come from?

d) What do you think has made it appear?

e) Where do you think it goes now?

f) Do you think you can make it go away?

g) Have you noticed anything like this happening anywhere else? (car exhaust, plane vapour trails, clouds, fog, mist.)

Interviews - Evaporation/Condensation

1 Handprints

Kit: damp sponge/cloth
paper towels

Press your hand onto a wet cloth and make a handprint on a paper towel.

Wave the paper towel in the air.

Discussion

a) What happens to the handprint?

b) Where do you think the water has gone?

c) Do you think you could make the water come back?

d) How can you make the handprint last longer?

e) How can you make the handprint go more quickly?

2 A tin containing iced water

Kit: tins
water
ice

Put some ice and water into a tin.
Look carefully at the outside of the tin.

a) What do you think is on the outside of the tin?

b) Where do you think it has come from?

c) What do you think had made it appear?

d) Do you think you can make it go away again?

e) Where do you think it goes to?

f) Can you think of anywhere else you have seen this happen?

WORDS TO BE CLARIFIED IN INTERVIEW/DISCUSSION

Evaporation/Condensation

Evaporation

Condensation

Disappeared

Dissolved

Dry/dried

Melted

Disintegrated

Shrunk

Sunshine

Questions and Response Check-list

These checklists are to help you build up an impression of the ideas which your children hold. There is room for you to add any comments you wish to make as well as ticks/tallymarks. You might like to look at each child individually or just get an impression of how common different ideas are in your class (this is probably more feasible).

The questions are meant to be used informally, as appropriate, not worked through from beginning to end. Additionally, the responses are there as a guide for you to check off against, rather than a source of "How many of you think it's the water?" They are **not** there for you to feed them in as ideas. **You** want to know **your** children's ideas so listen out for them telling you.

These checklists will be a valuable source of information for you during the intervention period; it is worth keeping a note of the dates when you make any particularly relevant observations.

You will also find a sheet of words whose meanings need to be clarified. It's important to make sure what a child means when they say that something has disappeared, for example. The conventional meanings, and breadth of meaning, of words can't be assumed. Any further words which you feel should be on this list, please add them and share them at the meeting.

CLASSROOM INTERVENTION STRATEGIES

		VOCABULARY	GENERALISATIONS	MAKE THE IMPERCEPTIBLE PERCEPTIBLE	TRY OUT CHILDREN'S OWN IDEAS	TEST THE RIGHT IDEA ALONGSIDE THE CHILDRENS'
STRATEGIES		Collect examples & non-examples of the word. This may involve a literal collection of items or a range of different activities.	Use discussion as a means of finding out which events (if any) children see as related to each other or to past experiences and help children make cautious generalisations.	If the children's ideas are based on a change, which they can't observe happening, see if it's possible to make that change occur visibly.	Use children's ideas as starting points. Encourage them to test their ideas in a scientific manner.	It may be worth feeding in a suggestion based upon the scientific view if the children find their own ideas to be inadequate.
CHANGES	metals	rust - colour and corrosion metal dissolve	. what is a metal? . metals change in water . some metals change in water . the air changes the surface of many metals	Fine steel wool rusting with water on it.	How can you make something rust? How can you stop something rusting? What happens when something rusts? (steel wool will react most obviously)	. coat metal in vaseline/ school glue (something which can be removed to check the metal) . look for rusty things indoors/away from water.
	moulding/ rotting	mould rot	. food moulds/rots if it is left in the air for a long time. . mould comes from something in the air.	Time lapse photography	What makes something mould? How can you stop something moulding?	Keep in a bag with most of the air sucked out.
	desiccation	stale 'left out'	. something goes into the air . something goes from foods that are left in the air		What makes something go wrinkly? How can you stop something going wrinkly?	Rehydrate a dried sample (doesn't work if it is too dried up)
	cooking	rises	. heat changes food irreversibly.	Use an oven with a glass door.		
	burning	disappear melt	. heat changes things irreversibly			
EVAPORATION / CONDENSATION	pure water	disappear shrunk evaporate soak dry disintegrate	. water 'dries' into the air . different things can help the water evaporate, e.g. wind, heat. . what different forms can water take?		Where does the water go? What can make the water go?	What different forms can water take?
	solutions	dissolve melt		water atomiser	coffee and water go.	Weigh solute before and after evaporation.
	condensation	condensation mist (a good alternative to condensation) steam smoke	comes from something in the air.		water comes from inside the can	. breath on warm and cold mirrors. . put ice into polystyrene cup . put empty can into fridge and bring it out into the room. . dye the iced water. . cover the can. . wrap the can in cloths.
GROWTH	animal	grow growth (involves a expand change of mass, enlarge volume and shape)	growth is continuous (human hair and nails) food becomes part of us			
	plant	seed bean	all green plants need certain conditions to grow.	Measure a rapidly-growing species, e.g. maize, at the beginning and end of each day.	What do plants need to grow? How can you make seeds start growing?	Measure rapidly-growing species at beginning and end of each day. Grow same plant in different Grow different plants in same media.
	potato		Some vegetables change into new plants if they are kept for a long time. The food in the potato is used up as the plant grows.			
	egg	develop hatch		Monitor an egg's development through the shell - candling hen's egg or magnifying cabbage white butterfly egg.		